PRAISE FOR J. B. MAXWELL

J. B. Maxwell has taught me so much. I'm so grateful to have found his books.

— JOHN ARBUTUS CO-FOUNDER OF THE
SEASONED GARDENER PA

So happy I found this book!

— TYLER LAYNARD

J. B. Maxwell delivers another helpful and useful book that we can all learn from and come back to time and time again.

— NICOLE TALA OWNER OF THE GARDENING
STORE IN CENTRAL PA

NORTHEAST MEDICINAL PLANTS: FORAGING FROM YOUR BACKYARD HOMESTEAD

NATIVE HERBALIST'S GUIDE TO IDENTIFYING WILD HERBS FOR HEALTH AND WELLNESS

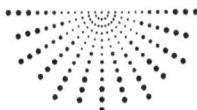

J. B. MAXWELL

CONTENTS

For my family which gives me immeasurable purpose and love every day

"The person who takes medicine must recover twice, once from the disease and once from the medicine."
–William Osler

INTRODUCTION

Did you know that nowadays, many people heal their bodies successfully by using herbal medicine?

Did you know that the famous herbalist Maria Treben purports to have cured many cancer patients by using herbs only?

In this book, we'll explore the various conditions that can be remedied by using medicinal plants. We will explore how this can improve your overall health and well-being holistically, but in a natural rather than a synthetic way. For some, medicinal plants may provide the long-term answers to pain or illness that you haven't been able to get from medical doctors. Not only that, but it's less expensive and can be less destructive to your body than traditional medicine.

If you're sick of being an experimental toy for Western Medicine, then taking natural healing into your own hands might be the answer. This can lead to a healthier family. It can also lead to the resolution of symptoms that recur and never seem to vanish. As a result, you could be living a healthier and happier life.

By being your doctor in the process of holistic healing, you'll find that you come to know your body better. You'll see what taking various herbs and plants does to your body, and by observing this, you'll end up more conscious of your bodily functions in general. As a result, it will be easier to make the right decisions for your health.

ABOUT THE AUTHOR

I am a family man who has taught himself how to be an expert in the field of medicinal plants. I grew up in a small farm town with no red lights in Maryland. Currently, I live in Pennsylvania with my beautiful wife and loving son. I love my family, our outdoor lifestyle, hiking, nature, gardening, and teaching others my knowledge of herbs. This knowledge extends to foraging wild herbs and plants, not just using them. Being outdoors makes me feel at peace.

I've always loved the idea of growing my own healthy, clean, organic food on a large enough scale that I can provide for my whole family without having to rely completely on grocery store produce as well as provide some for the neighbors. There's something about home-grown food's flavor that is unrivaled by its store-bought counterpart. On top of that, it's inexpensive and saves us a whole lot of money.

One of the things I enjoy about growing my food is that I have direct control over what is grown. I can ensure the plants are cared for properly while growing, which in turn makes high quality a certainty. This hands-on approach allows me to feed my family food with a high amount of nutrients, allowing for complete holistic health. I have been practicing homesteading principles for a little over eight years now on two properties, one being three-quarters of an acre and the other being an acre.

My knowledge, in turn, covers a large range of topics. This spans from the sphere of Western medicine to Chinese Eastern Medicine and Ayurvedic practices. I've traveled through the East where I gained this knowledge first-hand. Gaining this ancient wisdom gave me great comfort, and I am thankful for it.

Knowing this has allowed me to create a life that's off-grid with my family, and we are healthy and happy. I would love to share this knowledge so that others feel safe and empowered to do the same.

I love exploring forests and mountains with the knowledge that I'm safe and protected, whatever happens. This has led to a long-term journey of steady knowledge expansion about our natural environment. Because of my firsthand wilderness survival experience built up over more than a decade, I've reached a firm certainty of what I'm talking about. I've helped many people with herbal medicines and ancient practices that I've picked up along the way from shamans and practitioners of natural medicine.

I've studied and experimented with so much over the last 15 years. During this period, I have helped many people in America improve their lifestyles and get back into contact with Mother Nature. The result has been an improvement in their overall mental and physical health. I've gotten to the point where I have a deep passion to share all the knowledge and experience I've built up over the years.

My goal is to help you achieve a healthier lifestyle by helping you create a natural home apothecary. This is my passion because I know how much freedom, joy, and health improvement can be created thereby.

In only one book you'll learn about the most powerful plants you can use to optimize your health and promote your well-being. These plants have been selected carefully to cover the healing properties needed for the most common ailments. Even if you do not know herbal medicine and medicinal plants yet, this is a detailed guide for any kind of person.

1

WILDCRAFTING BASICS

*I*n this chapter, you'll be introduced to foraging. This will help you understand the importance and benefits of foraging to people's health. Wildcrafting can be a sustainable way of getting the herbs you need to stay vital in your everyday life if you practice it well. By following the right safety guidelines and adhering to ethical foraging standards, you'll find it to be a fulfilling addition to your life.

WHY WILDCRAFT MEDICINE FOR YOURSELF AND YOUR FAMILY?

One of the reasons wildcrafting is so important is that, with an added understanding of how much wild spaces can provide for you, you realize that you have tools at your disposal almost everywhere around you. These tools are nature's provisions for improving unwanted health conditions or for daily maintenance of general well-being. This will make you less dependent on other people when it comes to staying healthy, as you won't need to rely on pharmaceutical companies' doctors for all your conditions anymore.

The plants that you find in nature can help to propagate using proper wildcrafting techniques. Even when tubers are harvested, resulting in the whole plant being removed from the earth, you can still contribute to a healthier ecosystem. By removing the plant when it's seeding, you can plant the seeds in the earth you've just softened by removing the tuber. This causes the plant to maintain and often increase its numbers.

Besides the benefits the ecosystem receives from your responsible wildcrafting, there are many benefits that you'll receive too. The main one is that plants that grow in the wild often have more nutrients than those grown in agribusiness. You, therefore, benefit from increased vitamins and other nutrients in your diet.

You can help propagate the plants you find by using proper wildcrafting techniques. Even when tubers are harvested, resulting in the whole plant being removed from the earth, you can still contribute to a healthier ecosystem. By removing the plant when it's seeding, you can plant the seeds in the earth you've just softened by removing the tuber. This causes the plant to maintain and often increase its numbers.

In addition to the benefits the ecosystem receives from your responsible wildcrafting, there are many benefits that you'll receive too. The main one is that plants that grow in the wild often have more nutrients than those grown in agribusiness. You, therefore, benefit from increased amounts of vitamins, minerals, and fatty acids available in wild varieties of plants.

The mental effects of wildcrafting shouldn't be discounted either. By getting out in the open, breathing in the fresh air, and observing your environment, you'll find your mind becomes calmer. The hubbub you experience daily isn't hanging around with you out in the open, leaving you free to slow your mind down and notice the world around you.

ETHICAL FORAGING

There are many principles to take into account to ensure that your foraging is ethical and non-destructive. One of the most important is to ensure that you don't harvest a plant when there's only one clump of that plant in the foraging area, especially if it's a sparse clump. Ensuring that you only harvest when it's clear that the plant won't be eradicated from the site is the epitome of ethical foraging.

The takeaway with this principle is that when you're making use of wild spaces, only take what you need for your personal use. This way the space will be maintained and not made barren. Don't commodify the herb.

This goes hand in hand with the mindset that you're maintaining a relationship with the environment. There's a give and take, with you contributing to the propagation of plants by spreading their seeds and harvesting responsibly while the environment is contributing the materials you need for medicine, food, and various other needs. By finding time to go out foraging, you'll become more grounded, and you'll realize the true value of the land. Once you realize the value it holds, it will be very easy for you to respect the land and take steps, such as knowing the life cycles of the plants you'll be harvesting, to safeguard its natural resources.

Not altering the landscape of the wild space is an essential principle of ethical foraging. Keeping yourself from chopping down tree branches or driving off the road when possible will go a long way toward preserving the landscape. This can be difficult when you're trying to get to spaces that are untouched by pollution and chemical sprays. But, more often than not, you'll find that you can get to a truly wild spot while easily being conscientious about not damaging it.

A way you can make the environment better than before is to take a trash bag with you. This way you can pick up garbage you see lying around. In addition to helping keep the environment clean from others' trash, there's also a responsibility to pick up after yourself. If you've used something, keep it with you so that you can chuck it in the trash back home. You can go the extra mile by making it as if you've never been in space. This entails filling up holes you create and cleaning up any substances leftover from fires you make.

One of the more important points when it comes to ethical foraging is to ensure that you aren't breaking any laws. The law that would most directly affect you is trespassing. To avoid getting drawn into paying hefty fines for trespassing, ensure either that the area is for use by the general public or that you have permission from the private landowner to use their property.

SUSTAINABLE FORAGING TIPS

One way of avoiding sustainability issues is to harvest those plants that are considered to be weeds in the area. They may be considered weeds because they come from another area and pose a threat to local species. They also may be considered weeds because of an overabundance that threatens other species in the area. Harvesting these types of plants won't cause damage to the environment. Further, by keeping yourself aware of the at-risk species in your area, you'll be able to harvest other species that aren't listed. As a general rule of thumb, it's best to make use of the commonest wild plants in your area first.

Being aware of cultivation techniques can also be of use in ensuring sustainable foraging. In this way, you'll know how to sow the seeds of the plants you harvest. You'll also be aware of their life cycles so you can get them at the end of the cycle and facilitate their propagation. This is particularly useful to know for plants that are indigenous to the area you live in. Knowing this will enable you to cultivate the plants that belong in your region, thus preserving the natural landscape in question.

TOOLS

Tools improve safety and increase effectiveness while foraging. When using tools, you reduce unnecessary damage to plants while harvesting them. Below, I've delineated the best additions to any wildcrafter's toolkit. You can acquire most of these at a low cost, whether online or at a hiking store.

Pruning Shears

You use these to cut branches. They are extremely useful on most foraging trips. When choosing a pair of pruning shears, it's best to take a pair with non-slip handles or stippled handles. This prevents them from slipping when they're covered with plant juices. A sap groove on the front is a bonus because it lessens the amount of buildup while using your shears. As for what the shears are made from, I suggest a metal frame with a lightweight handle. This reduces the problem of snapping, chipping, and breakage that you may encounter with polymer shears. You'll also reduce unnecessary weight from a solid metal pair.

Knife

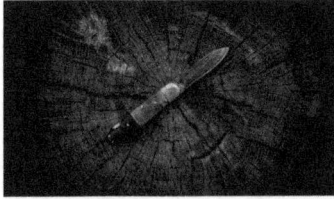

A knife is perfect for cutting off sections from tougher plants. There are curved-blade knives that are specifically designed for wildcrafting. They allow you to cut branches and twigs off in one motion. The best wildcrafting knives have rust-resistant blades and have non-slip handles. The rust-resistant blade is necessitated by the moisture present in plants you harvest. The non-slip handle is necessitated by the slippery oils and liquids that are exuded by plants that have been cut.

Scissors

Scissors are the best item to use when you're cutting off fine parts of a plant, such as leaves, flowers, or seed clusters. They can be used without causing any damage to the rest of the plant. I find that using a pair of kitchen scissors works quite fine because they tend to be rust-resistant, and they tend to be rather strong.

Foraging Bag

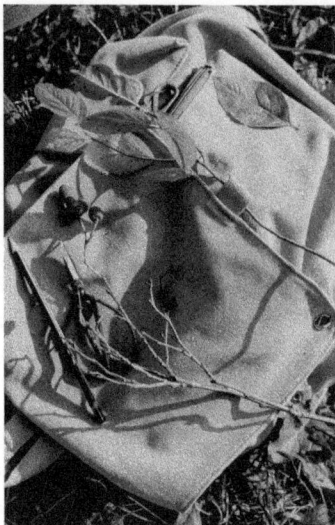

A bag is a necessity when it comes to foraging. A plastic bag is a no-no because it wilts your harvest. A paper bag is better because it provides more breathability. Some foragers even suggest keeping a bunch of paper bags in a small zipped plastic bag. This allows you to store a variety of plants that you've harvested. The weight of the paper bags is minimal, making any hiking you may do less tiring. The zipped plastic bag is useful for keeping your paper bags together and prevents them from getting wet before you've put them to use.

A step up from this is a canvas bag as it combines breathability, durability, and sustainability. A basket will also do the trick because it's sturdy and breathable. Your plants will thus remain fresh. The crème de la crème for holding your harvest is a bamboo bag. The bamboo fabric combines durability and breathability with easy carriage. This makes it perfect for harvesting trips.

Vegetable Brush

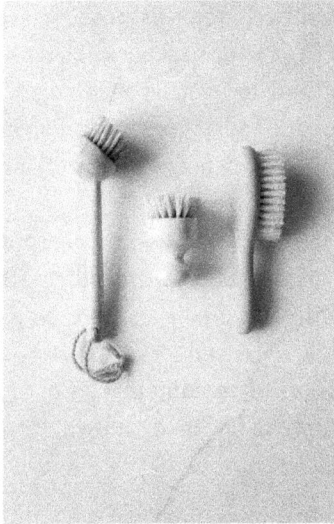

Vegetable brushes are a valuable addition to your toolkit because they allow you to clean dirt from plants that you've harvested. Dirt is an inevitable part of foraging, but much of it can be eliminated with a quick brush before placing it in your harvesting bag or basket. Not only does carrying a brush cut down on cleaning time at home, but the cost is also extremely low.

Shovel

Shovels make your wildcrafting adventures much easier. They allow you to get to roots and tubers with minimal effort. The best shovels for this purpose are ones that are small enough to fit in your backpack. They should preferably be rust-resistant and have sharpened edges. A heat-treated steel variety is great because it's strong. Polymer versions are often too soft, especially when dealing with hard, rocky soil.

Soil knife

Soil knives tend to be wide, with a sharp, smooth edge on one side and a serrated edge on the other. They are used for digging into hard-packed soil and can be great for cutting thick roots as well. Please go for a rust-resistant one because you'll find that roots often contain a lot of moisture and that soil can be rather wet a few inches down.

Magnifying Glass

When you're trying to identify a plant, sometimes you want to look at fine details. Occasionally your eyes won't be able to make out all the finer details, especially when you get a bit older. To get around this, you could use the zoom on your camera or phone, but this can have its shortcomings in terms of visual quality. A magnifying glass is much more effective for this purpose. You'll be able to zoom right in there without any reduction in visual quality. You can even get magnifying glasses that come in multiple strengths, allowing you to go to the most minute detail possible. With a trusty set of magnifying glasses, you'll be able to identify plants and any infections on them with ease.

Gloves

Gloves have the sole purpose of protecting your hands from sharp rocks and thorns as well as irritating hairs on the stems of plants. In the process of wearing them, they may become coated with sap or oils that contain poisonous or irritating substances. To prevent this from affecting your body, use care when handling gloves. Also, wash or replace your gloves occasionally. The best gloves also provide some level of breathability, allowing your hands to work without becoming hot and sweaty.

Guidebook

These are essential items for using in the field as well as for studying at home. It's best to have a variety of guidebooks because some books may have different pictures, illustrations, or descriptions than others. This helps when you're in the field because you'll have more than one perspective to determine which plant you're handling. The result is a heightened level of certainty. When you're not in the field, I advise going through your guidebooks in your spare time. An increase in theoretical understanding of the plant will help you work more smoothly when you're foraging in the wild.

SAFETY

Foraging in the wild is not without its risks. To prevent yourself from ending up in a sticky situation, use the below safety tips.

Go With Someone Else

If there are two or more people, there is always another set of hands in emergencies. Slipping on a rock and getting a concussion isn't something you plan for, but if it happens, it's far better to have someone there to contact emergency services. Foraging is one activity where the adage "too many cooks spoil the broth" doesn't apply.

Have a Mentor

Mentors give you the advantage of built-up experience. They can guide you when you have questions about what you've harvested, such as whether it's safe to eat. A mentor will sometimes be willing to come on foresting expeditions with you, thus providing a masterclass in wildcrafting. Take the advice of those with a lot of experience and use it to make yourself as skilled as possible in herb gathering.

Avoid Unhealthy Plants

Don't harvest unhealthy-looking plants. They might have infections, which isn't something you want to include in your dietary or medicinal stocks. Using a magnifying glass will occasionally be necessary for determining if something is wrong with the plant. But when you have determined that something is wrong, don't get yourself into trouble by including it with your edibles.

Avoid Spoiled Stock

Wild plants generally have a short shelf-life. Only harvesting what you need safeguards your health and protects the sustainability of the wild at the same time. To put it simply, why use old, ineffective herbs when it's so easy to get your hands on fresher ones?

Keep Potential Allergies in Mind

Concerning only taking the quantities you need, harvest even smaller amounts when it is your first time taking that particular plant with you. You may be allergic to the plant and not know it. Take the small amount to test it for an allergic reaction. You can harvest more once you've confirmed you don't have an allergy.

Be Mindful of Pollution and Pesticides

Some areas are not optimal for harvesting. These are generally areas that are close to sources of pollution, such as highways or smoggy city centers. It's also recommended that you avoid areas that are prone to being sprayed with pesticides. While some pesticides are plant-based and not full of synthetic chemicals, most aren't. Then again, even some plant-based pesticides can be harmful when used incorrectly, so altogether avoiding areas sprayed with pesticides is best. You could be putting yourself at risk of worse health by consuming herbs that have been sprayed with strong, harmful pesticides. Also realize that even if you wash plants that have been sprayed with potent pesticides, there may still be chemicals absorbed into the plant that you won't be able to wash out.

10 BENEFITS OF FORAGING WITH HERBAL MEDICINE

1. The plants are not harvested en masse. The result is a higher degree of sustainability because you

aren't purchasing your herbs from someone who harvesting them for commercial purposes.

2. Plants obtained in the wild normally have a stronger medicinal effect. Without as many contaminants farmed into them and without genetic modification, the herb remains naturally complete.

3. The herbs are fresher. When you purchase from a herbal retailer, they may already have stored the herb for a while. When you harvest it yourself, you have it from the moment it's been harvested, thus ensuring freshness.

4. You'll be saving money because you won't have to buy from a herbal store or online supplier. All you'll have to spend on is your transportation costs to reach your harvesting location and the initial costs of setting up your wildcrafting kit.

5. When you have taken the steps necessary to correctly identify a plant, you have certainty that you have the correct variety of it, especially if you've verified it with your mentor. The same can't always be said with purchased herbs. When herbs are purchased, you might occasionally find that the wrong variety has been identified. Foraging for the herb yourself, therefore, avoids this risk.

6. Having your own harvested herbs allows you the freedom to incorporate them into your daily life. When you purchase herbs, this could lead you to use them sparsely. Part of the reason this occurs is that you can't generally buy herbs in large quantities in herbal shops. You normally purchase a handful or so at a time, with pricing becoming unreasonable at larger quantities. Not only will you have access to larger quantities when you need

them while in the wild, but you'll also find that you have almost no hesitancy to use your stocks for general use. You know that you can just go harvest more when your stocks run low.

7. Self-empowerment is a by-product of foraging. By gaining the confidence to gather your medicinal and food provisions, you realize that you can become truly self-sufficient. This removes multiple barriers you might have encountered to taking higher levels of control of your life.

8. By foraging for your provisions, you have a direct say in what you use for your health maintenance and improvement purposes. When you purchase things from the pharmacy, they are regulated by the government, namely the Food and Drug Administration. While the FDA is there to safeguard the community, the products they approve can be far-removed from nature. The level of complexity these health products pose to someone that doesn't have lots of training on chemicals can be frustrating. This is all circumvented in general when you use herbal medicines that are closely connected to nature. You may have to resort to medical drugs on occasion, but these instances will be few and far between when you're able to resort to nature's provisions for general illness.

9. You will have a more complete say in what goes into your body. There are little to no additives in wild herbs, resulting in pure additions to your diet. The same can't be said when purchasing things from your grocer. Even if your grocer can provide good quality foodstuffs, you can't always obtain sourcing information. Further, you'll find that

often, the commercial needs of farmers result in pesticides and GMOs (genetically modified organisms) being used without your direct knowledge. This isn't the case with wild foodstuffs.

10. You will understand your responsibility to the world you live in. The correct term for this is 'stewardship,' and it refers to having a care-taking attitude toward nature and its inhabitants. By foraging, you build a relationship with Mother Nature, thereby increasing your level of understanding of the environment. The result is a heightened level of care toward the world in most things you do daily. This result often occurs unconsciously, being based on the goodwill you build up while foraging.

2

PLANT IDENTIFICATION

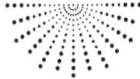

he main point of this chapter is to understand how
to identify a wild herb.

RECOGNIZING A PLANT

To effectively recognize a plant, you should be able to determine the types of leaves it has, the type of branches it has,
and the type of flowers it has.

Leaf Shapes and Arrangements

The purpose of the leaves of a plant is to help it get the
things it needs out of the air, to help it gather energy from
the light, and to help protect the plant. Leaves have multiple
shapes and arrangements. These vary from species to species,
regulating how sunlight is harnessed by the plant.

First, you should understand what the parts of a leaf
are. The image below should serve as a reference point.
The petiole is the sticklike structure connecting the base of
the leaf to the stem of the plant. The midrib is the single,
thick, straight, vein-like line stretching from the petiole to
the tip of the leaf. The midrib divides the leaf in two. The

veins stretch out from the midrib, supplying the leaf with water and nutrients. The leaf's margins are the outside edge of the leaf that gives it its shape. The part of the leaf between the petiole and the midrib is known as the base, whereas the part of the leaf furthest from the base is known as its tip. The final part to note is the blade. This is the part of the leaf that fills in the leaf between the veins and the margin.

[Note: some leaves don't have a petiole. In this case, you would say that the leaf is sessile]

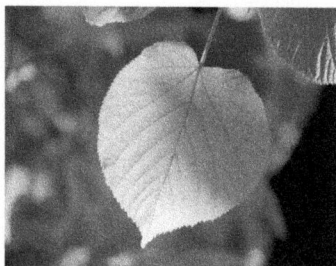

When it comes to classifying leaves, there are two main types. One is a simple leaf and the other is a compound leaf. A simple leaf is one leaf that isn't made up of smaller leaflets, whereas a compound leaf is made up of more than one leaflet coming from the same central vein (midrib).

Compound leaves are further divided into three sub-categories. The first is a palmate leaf. This is where there is more than one projection spread from a nexus, almost like how your fingers spread out from your palm. The other two types of compound leaves are a pinnate and a bi-pinnate leaf. A pinnate leaf is made up of multiple leaflets spreading out from a common stem. A bi-pinnate leaf is where smaller stems spring out of the common leaf stem, with each smaller stem sporting leaflets spreading out of it. When you look at a fern, you can see that there is a larger stem for each leaf. The larger stem then has smaller stems coming out of it. The leaflets of the fern then spread out from each smaller stem.

Arrangement

There are four main arrangements of leaves, namely opposite, alternate, rosette, and whorled. An opposite arrangement is where leaves come out on opposite sides of a stem, such as on an olive tree. An example of an alternate arrangement is a willow tree. With this type of arrangement, you'll notice that leaves grow on the sides across from each other, but none of them are directly opposite one another.

While not limited to one type of plant, rosette arrangements are commonly found in succulents such as aloe plants. In this arrangement, all the leaves of a plant come from one central stem. This is especially common on plants that don't grow very high. A whorl is also an arrangement of leaves from a central point, but rather than all the leaves of the plant coming from one central stem, there are many branches and twigs that each create a central point from which leaves spread. An example of this is a lemonwood tree.

The main purpose of all these arrangements is to optimize how sunlight falls on a given plant.

Shape

The shape of a leaf can help you identify a plant more accurately. The most basic distinction is whether a leaf has lobes or not. There are five other more precise shape classifications that I use when identifying a plant. They are linear, elliptical, ovate, truncate, and lanceolate.

A linear leaf shape is long and slim, like most grass leaves. Elliptical leaves are shaped like an oval, such as the leaves found on an American hornbeam tree. An ovate leaf is also in the shape of an oval. The difference between an ovate and an elliptical leaf is that the elliptic leaf remains the same width, but an ovate leaf is broader at the bottom and more narrow at the top. In other words, an ovate leaf is egg-like in shape. A further difference is that an elliptical leaf doesn't have lobes, whereas an ovate leaf could have lobes like an eggplant leaf does.

A lanceolate leaf can also be confused with elliptical and ovate leaves as it is also oval/egg-shaped. The difference is that a lanceolate leaf is broadest at the base, narrows at the midpoint, and then narrows to a fine tip. Lanceolate leaf types are much longer than they are wide, normally at least three times longer than their width. A willow tree leaf is an example of this.

The final shape I commonly use, a truncate leaf, is exactly what its name suggests. It seems to be truncated at the tip or the base, giving a flat appearance on one end. An example of this is a leaf on a tulip tree. The leaf has a few lobes along the side margins, but it seems to be flat at the base.

Margins

The margins of a leaf can vary a lot. That said, there are four main categories we use for speedy identification. There is a margin that is smooth the whole way around. Then there is a margin that is serrated the whole way around. The third is a lobed leaf, which has indentations; but the indents go less than halfway to the midrib. Finally, there is a parted leaf margin. This type of margin indents more than halfway to the midrib of the leaf.

Venation

Venation refers to the pattern or arrangement of the veins that can be seen on the surface of the leaf. There is parallel venation, in which veins normally occur parallel to each other in lines from the base of the leaf to the tip of the leaf. You also get dichotomous venation, in which the veins form a pattern that looks almost like a 'Y'. Palmate venation is when all the main veins come out from a central point on the leaf. And then there is pinnate venation, in which the veins are arranged coming out of the midrib.

Texture

Leaves of different species and subspecies have all sorts of textures on their margins that distinguish them from one another. This makes it much easier to identify a plant. Some of the most common leaf textures include wavy leaves, smooth leaves, toothed leaves, lobed leaves, and incised leaves. Both the textures and the degree to which these textures are present are distinguishing factors. For example, you can have a leaf with toothed margins where the toothing is fine, and another species where the toothing is ragged.

Other Factors

There are a few other distinguishing characteristics not mentioned yet. These include whether the leaves have thorns, hairs, or resin glands on them or not. The thickness of the leaf can also help in identifying the plant. The surface of the leaf, whether it be waxy, shiny, or dull, can also be used. The stiffness or limpness of the leaf also plays a role in identifying it.

Conifer Leaves

Conifers have slightly different types of leaves from other trees and plants, i.e. needles. These needles appear in multiple arrangements, depending on the species. These include clustered, bundled, single, linear, scale-like, and awl-shaped. Clustered, bundled, linear, and single should be self-explanatory. Scale-like refers to needles that are small and overlapping in nature. Awl-shaped needles are thin, linear, and taper to a sharp point.

Branching Patterns

There are multiple branching patterns you can use to identify plants, but the two most common types are opposite and alternate. Opposite branching patterns are where branches come out of a stem on opposite sides from each other and at the same point, sort of how ribs come out of your spine. Alternate branching patterns are where branches come out of a stem on opposing sides, but not at the same point.

One of the most useful things about using branching patterns to identify plants is that the branching pattern won't change from season to season. Even in winter, when many trees and shrubs have no leaves, you can still use their branching patterns for identification purposes.

Color and Number of Flower Petals

You may not always realize you're looking at a flower all the time because some of them look drab, very small, or like leaves. That said, you'll find that every plant has a flower of some sort. These may come out at different times of the year from plant to plant, depending on their life cycle, so expecting all flowers to bloom during the spring isn't going to provide you with the most accurate image of the plants you're foraging for.

The main characteristics you'll take into consideration when identifying a plant by its flowers include color, number of petals, size, petal stiffness, and petal/flower shape. Most field guides will give you a good description of the flowers of a plant, so it won't be difficult to determine if the plant you're looking at falls under the description in question.

There are often distinctive colors to the roots of a plant, its stem, and its leaves. Sometimes there are even distinctive colors for the same plant from season to season.

3

AGRIMONY

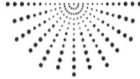

LATIN NAME

Agrimonia

DESCRIPTION

When they are young, agrimony plants have a basal rosette. As they grow older, their branches develop in an alternate arrangement. The leaves grow on petioles, and there are stipules at the base of the petiole. (Stipules are small leaf-like attachments at the base of the petiole). The leaf itself is a blade leaf. It's pinnate, with three to six pairs of leaflets on each leaf. It culminates in a terminal leaflet. There are small leaflets between each pair of larger leaflets. The leaf has a thin base that widens at the midpoint and then tapers to a point at the top. The margins are serrated, but the serration is large, rather than fine. The bottoms of the leaflets are hairy and have a gray tinge.

The flowers grow on sticks, creating structures called inflorescences. There are usually many flowers on each inflorescence, and they have an aromatic fragrance. The flower has five yellow petals that usually have a rounded tip. The sepals (the tougher petals that enclose the flower while it's still budding) also have five lobes, and they have small hooked spines underneath. There are between 10 and 12 stamens (the pollen-bearing structures in the middle of the flower), and two pistils (the parts in the middle of the flower that catch pollen in order to reproduce).

The fruit is small and has a green color that transitions to reddish color, then finally dulls to a brownish color. It also has many burr-like hairs that can cling onto your clothing. These burrs also go from green to red to a dull brown. In terms of size, the fruit doesn't grow much larger than a few millimeters. The shape of each fruit is bell-like at the base, ending in a small mass of burrs coming out the mid-points, and culminating in a rounded bud tip.

HABITAT

You can find agrimony on forest margins, in coppices, on banks of earth, in pastures, and in dryish meadows.

They should be in direct sunlight optimally, but they can also stand partial shade.

They prefer alkaline soil rich in calcium, growing particularly well in wettish, marshy conditions. You can also find them in dry soil; however, they normally don't flower in dryer soil.

SEASON TO GATHER THE PLANT

In the middle of summer, you'll find that it's best to harvest. If it's in bloom, this is an especially good time to gather some agrimony.

PARTS OF THE PLANT TO USE AS MEDICINE

The leaves are very useful for home remedies, so these can be harvested. But due to the prevalence of the plant and the number of ways in which you can use it, you can harvest the whole plant and hang it up to dry. This will allow you to use the entire plant for various reasons.

BENEFITS AND PROPERTIES

Agrimony has a host of benefits that you should be aware of. It can help cleanse the body. It can also help cleanse the mouth, cleanse mouth sores, clean out wounds, and clear up the skin. Due to its astringent qualities (helping to contract skin cells), it's not only effective for clearing up the skin, but is also effective as a lotion and as an ointment that heals skin irritations and sores. In the process of healing the wound, it will also help ease the pain you may feel in the wounded area, and it can help stanch any bleeding. As for your body's organs, agrimony can boost your kidneys and your liver and can ease stomach problems.

MEDICINE PREPARATION

Lotions and other skin applications: The easiest way to use agrimony for skin application purposes is simply to crush the parts of the plant you've selected so that the juice comes out. Once you have enough juice (about one tablespoon for every tablespoon of cream base), mix it in with a natural cream base, a natural fat, or a natural oil until it's mixed in smoothly.

Tea: made by adding a teaspoon of dried agrimony (any part of the plant that's been dried and crushed or cut up) to a cup of boiling water. You can steep for a few minutes, at which point it will be ready to consume.

Mouth gargle: made the same way as the tea, but you leave it to cool down afterward. I prefer to leave the agrimony in the water so that it gets stronger as I leave it overnight. Then a thorough swirl around the mouth will give the desired effect.

USING THE HERB

Lotions and skin applications: Apply directly to the affected area if the purpose of the application is to heal. You can also apply it to an open wound, so long as the other ingredients are safe for open wound application.

Tea: You should be able to drink this tea daily. Just note that it can have a constipating effect.

Mouth gargle: I prefer to let it get extra strong before using it. It has a more satisfying clean.

SIDE EFFECTS AND WARNINGS

Constipation is the most common side effect. Allergic reactions can also result from using this herb. Other side effects that can result from long-term use are nausea and vomiting. These reactions are due to the tannin content found in this herb.

FUN TIPS AND FACTS

Leaves of agrimony can be used as a source of yellow pigment.

AUTHOR'S PERSONAL STORY

I've helped many people cure digestive issues with agrimony. It helps even with IBS!

4
ALDER

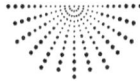

LATIN NAME

A lnus

DESCRIPTION

There is a large variety of alder species, with a vast difference in the sizes of alder trees, depending on the species. The dwarf species grows up to 15 feet tall, whereas some of the larger species grow up to 100 feet high. Most of the species, however, grow between 20 and 50 feet high. The tree is often used for wood. The wood is rather light, and the bark is quite thin. On top of the thin bark, you'll often find patches of lichen and moss.

The bark tends to be gray–sometimes lighter in hue, and sometimes darker. The bark itself also has unique coloration. There are two layers of bark on the tree. The outer layer is silvery gray, and the inner is brown. Once you've cut to the inner bark, it'll turn an orange color.

The leaves are arranged alternately on their twigs. Alder leaves have a serrated edge and are pointed at both the base and the tip. There's a slightly hairy texture underneath. Alder leaves are deciduous. Their leaves are green in the spring and summer, then in fall, the leaves go yellow, orange, red, and rust-colored. In some cases, the leaves are even purple. There is a distinctive, pleasant smell to this tree–similar to the scent of a cottonwood tree.

The flowers on alder grow on catkins (a type of inflorescence). The flowers are yellow at first then, after a while, they transition to a dark red color. The female flowers, on the other hand, develop on green growths, developing small red petals. Interestingly, most alder species first grow their flowers before they grow their leaves. The catkins give the tree a reddish hue in early spring.

Alder fruit comes in the form of half-inch wide cones. These start off green, then turn brown. Between the scales of the cones, you'll find small winged nuts.

HABITAT

These are very hardy trees. They can grow in areas that aren't fertile and that have recently gone through disasters such as fires or landslides or the clearing of land. Soil that isn't very fertile will increase in fertility when alder trees are growing in the area. This is because these trees can work symbiotically with bacteria, resulting in increases in nutrients present in the soil. Further, they can take the nitrogen from the air and transfer it to the soil.

Although they can grow in very rough areas, their ideal habitat is in areas with moist soil. They can often be found next to rivers and streams and in wetlands.

SEASON TO GATHER THE PLANT

Leaf buds should be harvested in February or March. Mature leaves, on the other hand, can be harvested from any time between the start of spring and the end of summer. The catkins can be harvested from when they appear until they become brown and hard in the winter months. As for the bark, this is best harvested from spring to fall.

PARTS OF THE PLANT TO USE AS MEDICINE

You can make use of the leaves, leaf buds, catkins (both mature and immature), and the bark. The bark is used more than the other parts of the tree because it's effective for medicinal purposes. I like using leaf buds and green catkins rather than mature leaves and mature catkins. They're just easier to work with and make a nice, fresh addition to your medicinal stocks.

BENEFITS AND PROPERTIES

Topical application is useful for the skin and for wounds. In terms of the skin, you'll be able to reduce acne breakouts, treat boils, and make your skin tauter. You can also eradicate inflammation and get rid of scabies.

For wounds, apply it either as a pulp, in a compress, as a wash, or in an ointment. The wash can be used to clean out a wound by getting rid of bacteria. Washes are also good for controlling bleeding and reducing swelling.

Sore muscles can be soothed greatly by applying alder creams or rubs. Rub it in quite thoroughly to soothe the muscles,

Your digestive tract will benefit greatly from using alder. Intestines will be stimulated, constipation will be alleviated, liver function will increase, and you'll produce more bile. The result will be more effective processing of food, such as better digestion of fats and better elimination of waste products.

Note that alder is said to have anti-carcinogenic attributes.

MEDICINE PREPARATION

Aperitif: Prepare the aperitif by getting some wine or fortified wine (white, pink, or red will do). Mix your alder (preferably fresh alder) with any other herbs you want to use. Don't worry if you crush it a bit–this will release some of the juices. Then pour over the wine (and ice if you prefer it cold). Leave it to infuse into the wine for a few minutes, at which point you should be able to sip and enjoy.

Poultice: Crushing the leaves or buds is all that's required to make an effective poultice.

Tea: Place the alder that you've harvested in a cup or mug–you won't require more than a teaspoon. Pour boiling water over it and leave it to soak for a few minutes. You can then strain and drink.

USING THE HERB

Aperitif: Use this before a meal as the purpose of an aperitif is to stimulate the digestive system.

Poultice: Once crushed, apply it directly to the irritated area.

Tea: Tea can be used for drinking, which is the usual reason for taking it. You can also use it for a mouthwash or a gargle to kill bacteria. It can also be used externally. For this purpose, you simply soak a cloth in the tea once it's cooled down a bit, then drape the cloth over any part of the body that you would like to soothe.

SIDE EFFECTS AND WARNINGS

Allergies are the most common side effect. This is especially so with alder because of the amount of pollen given off by the catkins blowing through the air.

A laxative effect is also a common side effect. When you've had a large amount of alder, you'll get a similar bowel reaction as when you take a mild laxative. Diarrhea is an indication that you should cut back on alder.

When you use bark from a tree within the first year of its growth, you could experience vomiting. The immature bark doesn't have the same properties as the older bark.

Other potential side effects you may notice are heart problems, bloody urine, an irritated stomach, weakening muscles, blood problems, and low potassium. Alder buckthorn in particular can have the side effect of cramps in your body.

FUN TIPS AND FACTS

The improvement in soil quality that alder brings about is due to the relationship it has with its symbiotic bacteria. The bacteria stimulate the roots so that they transfer nitrogen from the air into the earth. The tree then produces some natural sugars, which the bacteria consume.

AUTHOR'S PERSONAL STORY

When my wisdom teeth were growing, I was in really bad pain. Within 24 hours, gargling with alder alleviated my pain.

5

BEECH

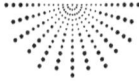

LATIN NAME

Fagus

DESCRIPTION

The bark is gray with cracks or fissures arranged horizontally on the tree. They tend to be tall, normally between 40 and 60 feet in height. They also have large, dense crowns.

This tree is characterized by an alternate leaf arrangement. The leaves themselves are simple, not complex. They have an ovate shape, but with a pointed tip and have margins that are characterized by fine teeth. The leaves are shiny yet have fine hair when they're still young. The leaves are rather large, with a width of approximately four inches, and a length between two and six inches. Venation is pinnate, with the veins coming out of the midrib coming to distinctive points, thereby accentuating the toothed effect on its margins.

Their male flowers are found on catkins, and their female flowers grow in pairs on stalks above the male catkins. The flowers, in general, have a yellow-green appearance and are small clusters of a bunch of fuzzy pollen-carrying filaments. They weigh down the stem holding the cluster.

A beech tree produces beechnuts, which have three sides. They're brown and found in a cupule with a spined surface.

HABITAT

Fertile soil with good drainage is ideal for this tree. They do well in direct sun and in shade. You'll find them mainly in temperate areas, which includes most of the East US.

SEASON TO GATHER THE PLANT

You can harvest from this tree between spring and autumn. Leaf sprouts normally appear in April and May, and the fruit normally appears in fall. Sometimes you can find shoots that haven't sprouted yet in February already.

PARTS OF THE PLANT TO USE AS MEDICINE

The bark, buds, leaves, and flowers can all be used medicinally.

BENEFITS AND PROPERTIES

Beech can be advantageous for your skin as it can help with, open pores, and skin issues. Piles can also be handled effectively by using beech products. It is an astringent, so it makes for good creams. Your digestive tract can be stimulated by using beech, improving your nutrient absorption. Furthermore, it improves kidney functioning, which makes waste elimination operate optimally.

You can also use it orally. When used in this manner, it can have an antiseptic influence and alleviate toothache.

MEDICINE PREPARATION

Tar: You can collect tar by heating the wood or roots. The liquid that drips out is tar.

Decoction: Simply mash up the leaves, buds, or flowers. The seeds are also particularly useful for this purpose. Boil them up in some water and strain out the mashed bits.

Poultice: Mash up the plant material to create a soft mush.

Tea: The leaves are best for making tea. Pour boiling water over them and leave them to soak for a few minutes.

USING THE HERB

Tar: Tar is wonderful for skincare, particularly as an antiseptic. You simply rub it onto any affected area of your skin.

Decoction: Use the decoction for internal use, especially for improving the functioning of your kidneys.

Poultice: Apply it directly to your skin. You can also place it on your head, forehead, or temples to ease headaches.

Tea: The tea is great for intestinal and digestive troubles.

SIDE EFFECTS AND WARNINGS

The nuts should be eaten in moderation because they are poisonous in large doses and can cause problems, especially in your stomach. This effect is largely avoided by cooking the nuts before eating them, as this eliminates a lot of the poisonous substance found in the raw nut.

FUN TIPS AND FACTS

Beech is associated with femininity and is often considered the queen of British trees, where oak is the king. In Celtic mythology, Fagus was the god of beech trees. Beech was believed to have medicinal properties. One of these was relieving swelling by boiling its leaves to make a poultice.

AUTHOR'S PERSONAL STORY

I had a patient who was losing hair. I told him to get some fresh beech and to create an extract to put on his scalp. He wouldn't believe it, but within a few months, he had his hair back.

6

BIRCH

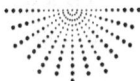

LATIN NAME

*B*etula

DESCRIPTION

These are rather large species of trees. They generally grow between 40 and 70 feet high with a canopy between 30 and 60 feet wide. Dwarf birch, on the other hand, is generally shorter than 30 feet high, with most being much smaller than that.

The bark is papery and easily peels off from the inner cork layer. The outer layer can be likened to paper. In addition to its distinctive peeling nature, it can also be recognized by its horizontal diamond-shaped marks. You can even remove the outer bark without damaging the cork at all, but some training and knowledge of how to do this are required first.

Birch has an alternate leaf pattern. The leaves on a birch tree are egg-shaped and have pointed tips. Some birch species have more of a wedge shape to their leaves. The margins are serrated, and in a few of the species, the teeth are quite conspicuous. Their surface is glossy in appearance. In terms of size, they're between two and three inches long. When looking at the tree from a distance, you'll see that the leaves are denser close to the top and that they often aren't present on the lower parts of the tree.

Birch trees also have catkins. These have small, yellow flowers. Male catkins are just over an inch long, and female catkins are about half an inch long. You'll find the catkins in solitary and clustered arrangements at the end of a peduncle, but clusters of three catkins tend to be the most common. They start flowering in spring during the same season when birch leaves bud. As the seasons wear on, these catkins transition from yellow, through red, to a reddish-brown color.

Small winged nuts can be found in the hanging catkins after they've flowered.

HABITAT

Birch trees prefer moist soil to dry or wet soil. The soil should preferably be sandy and contain a large amount of decomposed plant matter for extra nutrition. They do, however, grow throughout the US, so you may find them in different habitats that aren't described here.

SEASON TO GATHER THE PLANT

It's best to forage birch during the spring months. At this stage, the birch will be full of shoots, and the bark will be flowing with nutrient-rich sap. During this season, there will also be many young and supple twigs. You can also harvest birch leaves, flowers, and bark in summer and up into early fall on occasion. As for the fruit, this is most easily available at the end of summer when the cones aren't too hard yet.

PARTS OF THE PLANT TO USE AS MEDICINE

You will be able to use the leaves, twigs, catkins, and bark of the tree medicinally. The sap in the tree can also be tapped and used medicinally. Fungi growing on the tree can also often be used for medicinal purposes. Although not part of the tree, these fungi are commonly found on a birch tree. Remember to identify the fungi before use to make sure they are safe.

BENEFITS AND PROPERTIES

If you have issues with your skin, birch is the right tree for you to be foraging. The bark of a birch tree contains a large amount of a chemical called betulin, which is fantastic for skin healing and maintaining healthy skin.

Pain is something that can be alleviated quite well by the tree as well. This includes joint pain, pain from injuries such as breaking a bone, and pain in your muscles. This is mainly due to a chemical found in birch that's similar to aspirin.

Your internal organs and systems can benefit from ingesting birch. Things such as bladder infections, bladder stones, and kidney stones are commonly remedied when using birch.

MEDICINE PREPARATION

Tea (made with twigs): The tea you can make from birch isn't only good at healing and maintaining your health, it's also tasty. Pour some water that's close to boiling over birch twigs for the best tea, and leave it to brew for a few minutes. Younger twigs that are more pliable are best as the older, more brittle twigs have fewer nutrients.

Flour: You can use the dried and crushed inner bark in the flour you use for your everyday baking. To do this, crush the bark until it's extra fine because it's unpleasant to have to chew on tough, barky grains. It doesn't rise very well, so it's better to use with unleavened baked goods. The flavor is slightly bitter, but this can be satisfying in some types of cookies and bread. I mix it in with other flours I use, such as coconut flour, to create a blend.

Oil: Cover birch leaves, bark, catkins, or twigs with a carrier oil of your choosing in a mason jar. Place the mason jar in a pot of water, making sure it doesn't fall over. At this point, put on the stove and put on very low heat (too hot for your hand to be comfortable, but not so hot that the jar would burst or so hot that the water boils). Keep the pot filled with water and on the stove for the entire day, and repeat the same process for the next two days as well. At this point, birch should be thoroughly infused into the oil. You can also use a process where you don't use the stove (which is the process I use because I don't like wasting gas unnecessarily). I place the birch pieces in a jar of carrier oil and put them outside in the sun during the day. I repeat this process for about five days, at which point it's ready to be strained and used.

Washes: simply soak birch pieces, whether it be leaves, bark, twigs, or flowers, in some water for a few hours. You can then use it to cleanse your skin. When injured, you can use this as a soak for a cloth which you then drape over the injured area. And when it comes to your hands and feet, you can use it as a soak to relieve any pain they may be going through. It's a great soak for your feet because it doesn't only relieve pain, but it can also be used to assist with killing foot infections and fungi.

USING THE HERB

Tea: The tea is rich in vitamin C and has a flavor similar to peppermint tea. I find that it's unnecessary to add anything to it to adjust the flavor, so I tend to not use honey or other herbs and spices in this type of tea. That said, it can be rather pleasant to add catkins to your tea for a fuller flavor.

Flour: I sift the bark through a fine strainer after grinding it because extremely fine bark is the easiest to work with.

Oil: Use the oil for any number of skin conditions–such as acne, eczema, or dry skin–and pain relief in massages. The oil is also particularly useful for making herbal creams and ointments. It makes for particularly soothing skincare.

Washes: Pouring the water you've infused as a wash into your bath water can also be beneficial. When you have sore muscles, you can do this to make a relaxing body soak.

SIDE EFFECTS AND WARNINGS

The pollen given off by birch trees is a very common allergen. For this reason, if you go foraging for birch in the spring, and you know you're allergic to its pollen, take some antihistamine when you go. This way, you can harvest without being affected by all the pollen in the air.

FUN TIPS AND FACTS

Birch trees symbolized renewal, purification, and beginnings in Celtic mythology.

AUTHOR'S PERSONAL STORY

Whenever I stand in front of a birch tree, I feel so much purity. As I told you in my other book, the first time I ever drank birch sap, I was fascinated by the immediate feeling of purification that flowed into my veins.

7

BLACKBERRY

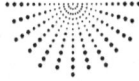

LATIN NAME

*R*ubus

DESCRIPTION

A blackberry bush can grow to a height of a few feet, but it spreads out well. It can grow to a 20-foot-long mass, and in some areas, it forms a large thicket with more than one blueberry bush amassed into one. Their roots grow quite deep for a bramble bush–approximately three feet deep. This, in combination with their prickles, makes blackberries very sturdy in the ground.

Blackberries come in the form of bushes with a lot of prickles. From afar, they may just look like a messy mass of prickly branches and leaves. And in the winter, they may just appear to be a dried, prickly mass of branches. The stem and branches of the blackberry are woody and covered in prickles. They are rather long, with some being as long as 20 feet.

They tend to arch up and have a green, red, or purple color. The prickles along the stems can vary in size, some being rather large, and others appearing as small hairs.

Blackberry bushes have alternate leaf patterns. Each leaf is made up of three to five leaflets that are one to three inches long each. This terminal leaflet (the one forming the end of the stem holding a cluster of leaflets) has a three-lobed shape. A few of the species don't have a three-lobed terminal leaf, but rather one lobe. The margins of these leaflets are normally deeply toothed, although some blackberry species have a finer toothed pattern on the margins. As for the design of each leaf, you'll notice that there are prickly hairs on the veins and on the stems to which the leaflets are attached.

At the base of the stems that support these leaves, you'll find a pair of stipules (small leaflike structures). The leaves are normally a darker green on the top and a lighter green below. The lower side of the leaf can get so light that it has a white-green appearance.

HABITAT

They are found in all types of settings, but the most common is in a forest or wooded area, or in areas where the ground has been cleared, such as in a landslide. That said, they can be found almost anywhere once they've been exposed to that area.

SEASON TO GATHER THE PLANT

The flowers can be harvested as soon as they start appearing. They generally bloom from April to May, and transition to fruiting at the end of May. The fruit is best harvested when they are ripe, which will be when they've gained their dark purple-black color. This is usually from July to August. The leaves can be harvested in spring and summer, but it's best to get them before they're dark and mature. When harvesting the leaves, using gloves is best because you can prick your fingers if you don't use them.

PARTS OF THE PLANT TO USE AS MEDICINE

The fruit, the flowers, and the leaves can be used medicinally.

BENEFITS AND PROPERTIES

Blackberry has some stellar benefits when it comes to keeping your blood in good condition. If you have anemia, this will help because blackberries help to absorb iron. It can also regulate a menstrual cycle better, reducing the amount of bleeding you experience. Your digestive and excretory systems will also benefit. If you have trouble with diabetes, blackberry leaves can assist with lowering your blood sugar. Diarrhea can be handled by taking blackberries. Likewise, if you have dysentery or cholera, blackberry leaves will help too.

Your oral health can also be improved by blackberries. Open sores in your mouth can make eating and speaking less pleasant to do. Blackberry juice can help with this. Other oral issues that can be improved with the juice include a sore throat and inflammation in your gums.

If you're feeling under the weather, blackberry can provide you with a solution. Blackberry leaves can strengthen your immunity. This, in turn, can assist with breaking fevers and getting over colds.

A final consideration is that blackberry has been linked with some people having successfully overcome cancer.

MEDICINE PREPARATION

Tea: for blackberry tea, you'll need about a teaspoon of dried blackberry leaf pieces. You throw about a cup of boiling water on this and leave it to soak for five minutes. After five minutes (or longer if you prefer a stronger brew), it'll be ready to drink.

Decoction: a decoction will require a handful of dried blueberry leaves, flowers, or fruit. Put this in a small pot and bring it to a simmer on the stove. After half an hour of simmering, the decoction will be ready for use.

Tincture: the tincture will require that you fill a sealable jar with fresh blackberry leaves or berries and cover it with vodka or brandy. You seal off the jar and leave it somewhere (not in direct sun) to soak for four to six weeks, giving it a shake every few days. After this, strain out the leaves and refrigerate the tincture. It will remain good to use for more than a year.

Poultice: thoroughly mix in a tablespoon of infusion or decoction, or a teaspoon of tincture, with three tablespoons of coconut oil.

USING THE HERB

Tea: you can have tea up to three times a day.

Decoction: the decoction is ideal for mouth infections, mouth sores, and a sore throat. Simply gargle it or use it as a mouthwash, and you'll start feeling the benefits. It's also good for drinking, especially if you want to handle a runny stomach.

Tincture: taking a teaspoon at a time will give you a strong enough dose. It's especially effective for coughs and a sore throat.

Poultice: the poultice is great for remedying scaly skin.

SIDE EFFECTS AND WARNINGS

If your stomach tends to be sensitive, avoid drinking a blackberry product. It contains tannins, which can result in nausea on occasion. This is especially the case when using blackberry fruit rather than leaves or flowers.

FUN TIPS AND FACTS

Just like its sister fruit, the raspberry, the blackberry enjoyed pride of place in Greek mythology. It was said that it sprang from the bloodshed by Titans in their wars against the gods.

AUTHOR'S PERSONAL STORY

I've always loved going into my grandparents' garden and eating a handful of blackberries with sweat on my body, sun in my face, and wind blowing through my hair. Such a refreshing and nourishing fruit!

8

BLACK CHERRY

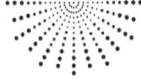

LATIN NAME

*P**runus serotina*

DESCRIPTION

The black cherry tree grows to a moderate height of 20 to 30 feet. The bark is gray, brown, or black, and it has a scaly texture. The bark turns up at the edges. Under the bark, there is a greenish layer that later changes to a creamy color with a green undertone. The branches growing out are green when younger and red-brown when older. The inner branch of the tree gives off an almond smell.

Leaves on this plant grow in an alternate arrangement. They're simple leaves that grow between two and six inches long. Their shape is lanceolate or ovate, and their tips are pointed. The margins of these leaves are finely toothed, with the teeth curving inward towards the tip. The leaves are green and shiny, and the lower surface is paler than the upper. In the fall, these leaves go yellow or orange. The leaves are on petioles, and they have glands at the base on either side of the petiole.

Flowers grow as white blossoms on a long inflorescence and smell sweet.

The fruit is a single drupe with a large pip inside. It's red when it's not ripe yet, and it's dark red or black, and it has a tart, bitter flavor when it's ripe.

HABITAT

Direct sunlight and chalky or limestone-rich soil is the ideal habitat for black cherries. It's normally found in wooded areas, especially on sloping or hilly wooded areas.

SEASON TO GATHER THE PLANT

April to June is when the flowers bloom if you wish to harvest the flowers. August to September is when it begins to seed, so this is the best time to gather the fruit.

PARTS OF THE PLANT TO USE AS MEDICINE

The fruit and the bark are both good for medicinal use.

BENEFITS AND PROPERTIES

Black cherry helps with muscle healing, especially after you've done heavy work or strenuous exercise. Further, it has a sedative effect, which can help with relaxing muscles. If you're having spasms in your muscles, you'll find black cherry can make it subside.

If you have trouble falling asleep, this can help you overcome your insomnia.

Issues related to the nasal area and the immune system can be overcome with black cherry. The benefits include reduction of coughing, drying out mucus, and soothing a sore throat. You can also use it to treat colds, fevers, and whooping cough.

Black cherry works wonders when used on the skin. It can firm out the skin, and it can dry out things oozing out of the skin, making it great for wound healing. Further, it reduces inflammation–though this isn't restricted to topical inflammation.

Black cherries are also beneficial to the digestive system. It has a stimulating effect on your gut, allowing your food to process better. When experiencing diarrhea, you can take it to slow down and stop the condition.

If you experience pain, this is one of the better herbal shrubs that you can use. It can be used to soothe pain both internally and topically. Gout is one example of a condition that can be alleviated by using it topically.

Lung issues can also be treated with this shrub. Bronchitis and asthma can both be treated with it.

Finally, you'll find that it can boost the heart overall. Your overall blood circulation can also be improved using it.

MEDICINE PREPARATION

Juice: Pit the black cherries and juice them. Add mint or other natural forms of flavor if the cherry juice is too sour or tart for your liking.

Tea: Dry and powder the bark. Add a teaspoon of this to a cup of boiling water and seep it for ten minutes. You can sweeten it or add natural flavor enhancers–such as pieces of fruit.

Tincture: You'll need an ounce of dried and ground bark, 16 ounces of vodka, and a large mason jar. Put the bark in the jar then add the vodka. Stir it together, seal the jar, and shake the jar. Leave it in a sunny place for a month to a month and a half. Shake it every few days. Strain out the bark and pour the tincture into dark glass bottles.

Ointment: for this, you'll need half a cup of the bark–dried out and crushed. Boil the bark in a saucepan filled with three cups of water for a few minutes. After this, bring it down to a simmer for approximately two hours. Remember to stir every fifteen minutes or so; otherwise, the bark might burn to the bottom of the pot. When the four hours are over, pour the remaining liquid through a fine strainer so that you don't get any annoying bark pieces in your ointment. Use about four tablespoons of the liquid and place it in a saucepan on very low heat. Place in two tablespoonfuls of beeswax, and two tablespoonfuls of coconut oil. Let these melt into the liquid, gently stirring it so that the ingredients combine. Pour it into containers of your choice and let it cool into a more solid form.

USING THE HERB

Juice: The juice won't last long, so consume it within a few days. You can have up to two tablespoons per day, preferably before bed. This will allow you to sleep better as it contains natural melatonin. Note: you can freeze it.

Tea: if you have a cough, take the tea three times a day.

Tincture: The tincture is great for improving digestion. You can have a half teaspoon daily, either directly or mixed in with water.

Ointment: This helps to tighten and firm up the skin.

SIDE EFFECTS AND WARNINGS

Allergic reactions can take place, especially when you're allergic to other fruits.

Don't use large quantities of black cherry as it can lead to toxicity in your body.

If you're on any type of medication, first consult with your healthcare provider before taking it. It produces interactions with several types of drugs.

Women who are pregnant or breastfeeding should avoid using this plant.

Ingesting the leaves can cause cyanide poisoning, so avoid using them.

FUN TIPS AND FACTS

American voodoo traditions have used cherries in love spells.

AUTHOR'S PERSONAL STORY

It's such an immune-boosting berry! It's rich in antioxidants and melatonin, which strengthen the immune system. I take it for illness prevention, and it always works! I haven't been sick in years because I bolster my health with herbal medicine.

9

BLUE VERVAIN

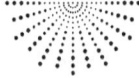

LATIN NAME

Verbena hastata

DESCRIPTION

If you take a look at what is going on underground, you'll find this plant produces rhizomes. The rhizomes allow the plant to spread through its direct area by sending out root systems horizontally that can then send up plant shoots. As a tip, if you want to grow blue vervain in your garden, you can dig up a rhizome in the wild to plant it at home.

The stem on this plant is hairy, and it has a square shape. It can also be either red or green in color. This plant grows between two and five feet tall.

The leaves are arranged, and they grow in pairs along the stems of the plant. The leaves are lanceolate with toothed margins, and they grow about six inches long and an inch wide. The flowers are purple-blue with lobed petals. They are clustered loosely along stems that are up to five inches long, and individual flowers are about a quarter-inch in diameter.

The fruit of this plant is a group of four red-brown, oblong nuts found inside the flower.

HABITAT

You'll find blue vervain in areas that have full or partial sun. They grow well in moist soil, so areas in which you'll find them include thickets, prairies, meadows, and along the sides of rivers.

SEASON TO GATHER THE PLANT

The flowers bloom in summer, making this the ideal time to harvest this plant.

PARTS OF THE PLANT TO USE AS MEDICINE

You can use the nut-like seeds of the plant as well as the roots, the leaves, and the flowers.

BENEFITS AND PROPERTIES

The topical application of this herb aids in treating cuts, sores, acne, and cramps.

A mental benefit you could experience by taking this plant is alleviation of depression.

Internal benefits include alleviation of cramps, as well as the remedying of ulcers, jaundice, headaches, fevers, and coughs.

There has also been some association between taking this herb and the reduction of tumors.

MEDICINE PREPARATION

Powder: Roast the seeds and ground them up.

Snuff: The flowers should be dried and crushed into a powder.

Essential oil: Use four handfuls of fresh blue vervain (cleaned well first). Put it in half a pot of boiling water and cover it with the pot's lid upside down (so that the evaporated oils condensate and drip back down into the pot). Simmer this until the water level has dropped to a quarter of the pot. Switch it off and let it cool. Place the pot in the refrigerator overnight and scoop off the filmy residue the following day. This filmy residue is the congealed essential oil. Put it in a dark glass bottle.

Infusion: Place a handful of fresh or dried blue vervain in a glass of water. Let it soak for four hours, then strain out the vervain. The liquid is now ready to drink.

Tincture: Fill a jar with cleaned, fresh vervain. Cover it with vodka and seal the jar. Put it in a dark, cool space for six weeks, shaking it every four days or so. Strain the tincture into a dark glass jar.

USING THE HERB

Powder: The powder can be mixed into flour. It can also be mixed into smoothies and juices.

Snuff: When experiencing a nosebleed, use the snuff to stem it.

Essential oil: This is powerful, so you should dilute it into water or oil. Don't use more than a few drops per application.

Infusion: Use it within a few days.

Tincture: It will last for years.

SIDE EFFECTS AND WARNINGS

It can exacerbate anemia, especially iron-deficiency anemia.

Interactions with drugs can occur when taking this herb, especially with hormone medication and blood pressure drugs.

If you have a large amount of blue vervain, you could experience diarrhea and vomiting.

FUN TIPS AND FACTS

In ancient Egyptian mythology, vervain was sacred to the goddess Isis. Whenever she cried, vervain grew where her tears hit the ground.

AUTHOR'S PERSONAL STORY

I had this patient once, an older woman. When she walked into my studio for the first time her breath was so heavy and fast, her voice was shaking, she seemed very confused and out of herself–she was suffering from anxiety!

From the moment I prescribed her blue vervain, she changed. On the next appointment, 1 month later, she seemed like another person; calmer, slower, more present, and focused.

10

BORAGE

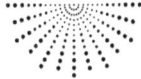

LATIN NAME

*B**orago officinalis*

DESCRIPTION

Borage is a plant that grows up to 36 inches tall and 24 inches wide. Its stems are hairy and prickly.

Borage leaves are between a rectangle or oval in shape and arranged alternately. Their margins are wavy, and their veins are visible on the surface of the leaf. They grow between two and six inches long, and y find there are short, stiff hairs that get quite prickly when mature growing on the leaves and stems. These hairs give a silver shimmer to the plant. In the fall, the leaves go brown and yellow, while in the summer and spring they're bright green.

The flowers are blue, but on occasion, you can find borage with white blossoms. They grow as drooping clusters on red stalks and have a star-like shape. The five-pointed petals are slightly hairy, and a short purple-black column protrudes from the center.

Its seeds grow above ground, allowing the plant to self-seed.

HABITAT

Borage likes direct sunlight and can tolerate slightly dry soil. It commonly grows in disturbed soil and on cultivated land.

SEASON TO GATHER THE PLANT

Spring and summer is normally the right time to harvest. When it's flowering, harvesting the flowers will keep the plant from deteriorating as quickly.

PARTS OF THE PLANT TO USE AS MEDICINE

Its flowers and leaves, as well as the oil from its seeds, are used as medicine

BENEFITS AND PROPERTIES

Borage can help purify your blood and prevent heart disease and strokes. You can also use it to treat adrenal deficiency, manage diabetes, regulate urine flow, and alleviate premenstrual syndrome.

In addition to all this, borage can increase your lung function, prevent lung inflammation, help manage asthma, and mitigate acute respiratory distress syndrome. You can use it to get over colds and bronchitis thanks to the way it promotes perspiration, and it treats coughs and fevers.

Topically, this herb can treat skin conditions including eczema, rashes, and itching. It can also reduce pain and inflammation in joints that have rheumatoid arthritis, and it can be used to reduce gum inflammation.

There are multiple other baby-related benefits to using borage like increased breast milk production. Additionally, can help premature infants grow and become stronger, and it aids in nervous system development

You can use it to manage stress and depression, and it can help with symptoms of ADHD (attention deficit-hyperactivity disorder). It may help manage alcoholism as well.

It also has a sedative effect, which is great if you have trouble sleeping.

MEDICINE PREPARATION

Oil: Cold pressing the seeds gives the highest quality results.

Tea: Use a quarter teaspoon of chopped-up fresh leaves or flowers. Put this in a cup of boiling water and leave it to steep for 10 minutes. Strain and drink.

Tincture: Fill a jar with fresh flowers and cover them with vodka. Seal the jar and put it in a dark space for a month. Strain out the flowers and pour the tincture into a dark glass bottle.

Infusion: Use a cup of water and a quarter cup of fresh, bruised leaves. Put the leaves in the water and leave it to steep for about four hours. Strain out the leaves, and it should be ready to use.

Poultice: Mash up some fresh borage and mix it in with a tiny bit of water and flour. It should be thick and sticky (the measurements of the ingredients aren't so important). Put it in a cloth and wrap the cloth around the affected body part.

USING THE HERB

Oil: This is good for external use (by rubbing it onto affected parts of your skin), as well as internal use (such as taking a tablespoonful).

Tea: The tea is great for stress.

Tincture: This is perfect to use for internal benefits.

Infusion: The infusion can be drunk, or you can use it as a wash.

Poultice: Use this for inflammation, soreness, bug bites, or stings.

SIDE EFFECTS AND WARNINGS

It's best not to use the oil for long periods. There are potential side effects that haven't been fully studied yet.

People can also experience allergic reactions when using borage.

When using borage while breastfeeding or on your baby, you'll need to consult a healthcare expert first. Some types of borage contain pyrrolizidine alkaloids, which are poisonous to humans.

Pyrrolizidine alkaloids damage the liver, so avoid borage if you have liver issues and you're not sure if the plant you have is a variety that doesn't contain these alkaloids. Further, avoid it if you're on any sort of liver medication, as there could be interactions.

Don't use borage if you bleed or bruise easily. It can exacerbate these issues. You should also avoid it if you're on medication that slows blood clotting.

Individuals taking phenothiazines should also avoid borage. Phenothiazines are included in drugs that prevent nausea and in some psychiatric drugs.

FUN TIPS AND FACTS

Borage is sometimes referred to as the 'herb of gladness'. The Roman scholar Pliny the Elder believed that borage was the 'Nepenthe' in Homer's Odyssey, which induced absolute forgetfulness when infused in wine.

AUTHOR'S PERSONAL STORY

This plant combined with a couple of others and a lifestyle transformation helped my mother a lot with depression during her menopause.

11

BURDOCK

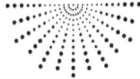

LATIN NAME

Arctium

DESCRIPTION

Burdock can grow rather large, reaching heights of 10 feet. It's a plant with a central stem that's fleshy and round and thinner stems coming out at various points.

The leaves are alternating, and there are large basal leaves, which can grow up to three feet wide, in the plant's first year. As it grows, large heart-shaped leaves with downy bottom surfaces grow from peduncles on the branches. The top surface of the leaves is green, and the bottom surface is whitish.

The plant has a combination of sweet (especially while flowering) and slightly bitter scents.

A tall flower stalk grows between three and six feet tall in the second year of growth, allowing a cluster of pinkish-purple tassel-like flowers that grow on top of spiked, circular seed pods. The flower heads are about an inch across. The spiky seed pod dries out when the flower dies and hooks onto animals or people that pass by. It can be quite irritating if it hooks onto your skin.

HABITAT

It likes soil that's rich in nitrogen, and it prefers full or partial sun. It grows well in disturbed areas.

SEASON TO GATHER THE PLANT

Between July and September is when to harvest the flower, and September to October is best for seed harvesting. The roots can be harvested in the fall without inflicting too much damage on the plant. The leaves can be harvested at any time. It's best to harvest from the plant during its second year (once the central flower stalk has shot up) because harvesting from it in the first year may prevent it from growing the stalk–it's a plant that lives for two years.

PARTS OF THE PLANT TO USE AS MEDICINE

The roots, seeds, leaves, and flowers can be used.

BENEFITS AND PROPERTIES

It's good at combating fungi, bacteria, blood impurities, toxicity, and fluid retention. It also assists with feeding the healthy bacteria in your body. You can use it to get over constipation, regulate urination better, lessen flatulence, and improve digestion.

Burdock is good for overall skin health. This herb can

help with inflammation as well as itchy and scaly skin. In addition, it can help regulate the sebaceous glands that produce oil, which is great for acne.

You can also use burdock for increasing bile production, increasing perspiration (and thereby preventing overheating), and encouraging the flow of lymph in your body.

MEDICINE PREPARATION

Tincture: Use two-thirds of a jar of chopped burdock root and cover it with vodka. Close the jar and put it in a cool, dark place for three months. Shake the jar occasionally until the three months come to an end. Strain the tincture into a dark glass bottle, and it'll be ready to use.

Tea: Use a tablespoon of the chopped fresh root, or two tablespoons of chopped and aged dried root. Boil three cups of water in a pot and put in the burdock. Let it simmer for half an hour, then steep for another 20 minutes once the stove is off. Strain out any of the roots, then drink the tea.

Oil: You'll need to break the seed heads and take out the seeds. Dry them and crush them. Use a ratio of one part seed powder and three parts extra virgin olive oil, placing it in a jar. It'll be ready to use instantly. Cover the jar with a cloth as the mixture needs to breathe. Store it in a refrigerator.

USING THE HERB

Tincture: It's very potent, so you should only have a few drops per day.

Tea: You can drink the whole pot in a day. Just make sure that you space it out throughout the day because having too much of it in one go will cause you to urinate a lot.

Oil: Use this for external purposes.

SIDE EFFECTS AND WARNINGS

Burdock can cause allergic reactions, and you should avoid high doses to prevent toxicity. Further, be extra certain when you identify it because it looks similar to belladonna nightshade, which has strong toxic effects.

Avoid using this plant during pregnancy and while breastfeeding. Further, avoid it while trying to get pregnant.

Avoid it when dehydrated, as it increases urination. Also, avoid it if you're on water pills, as it may make the water pills too strong.

FUN TIPS AND FACTS

Burdock extract can enhance sexual functioning and can increase sexual desire, so it can be used as an aphrodisiac.

AUTHOR'S PERSONAL STORY

It's crazy the amount and size of pimples people get when taking this plant's tincture! It's such a deep detox for the body.

CATNIP

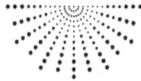

LATIN NAME

N epeta cataria

DESCRIPTION

Catnip plants can vary slightly in size from species to species, but at full size, most species grow up to three feet tall. You'll find that the stems tend to be thick with a hairy surface.

The leaves on a catnip plant tend to be toothed and hairy. The leaves have the shape of a heart-like oval, and while the leaves are silvery green for most catnip species, in some cases, the leaves can be dark green. The plant has a definitive minty smell to it. The flavor, on the other hand, is either grassy or woody.

The flowers occur in clusters on the stem, with each flower on the cluster having small tubular openings. Flower color is the main difference between the various catnip species. They're mainly white, but some have a violet tinge, and others are completely violet.

HABITAT

They can do well in soil with low nutrition. Gravelly or sandy soil is best. Sunlight should be direct, ideally. It grows in a wide array of areas, ranging from riverbanks to dumping grounds.

SEASON TO GATHER THE PLANT

It grows all year long, meaning that you'll be able to gather a harvest of the leaves and twigs the whole year round. The best season to harvest, however, is when the flowers are in bloom from July to October.

PARTS OF THE PLANT TO USE AS MEDICINE

The flowers, roots, and leaves can be used.

BENEFITS AND PROPERTIES

Catnip tea has a relaxing effect thanks to the chemical nepetalactone present in the plant. This chemical has a sedative effect when ingested, helping you with anxiety and insomnia. This is different in cases when the root is used, under which circumstances, catnip acts as a stimulant.

Your gastrointestinal system may benefit the catnip tea's ability to ease indigestion, gas, and stomach discomforts. Water retention is also reduced by taking catnip tea because there is a diuretic quality to the catnip herb. Similarly, it can help expel the placenta after giving birth as well as stimulate the uterus in cases of delayed menstrual cycles.

Other conditions that could be alleviated by catnip tea are fevers, hives, coughs, and viruses. In other words, there is an immune-system-boosting effect brought about by consuming the tea. Arthritis sufferers, be aware that this tea can aid in soothing this condition too.

MEDICINE PREPARATION

Tea: Use a sprig of catnip and put it in a cup of boiling water. Leave it to steep for 10 minutes before removing the catnip. The tea will now be ready to drink.

Tincture: Use a jar full of chopped-up fresh catnip and vodka. Put the catnip into a jar and cover it with vodka. Close the jar and place it in a cool, dark place. Leave it there for three months, giving it a shake weekly and adding a bit of vodka if the level has dropped. When the three months have passed by, you can strain the catnip out and place the tincture in a jar.

Infused oil: Finely chop up fresh catnip (enough to cover the bottom of a casserole dish) and place it on the bottom of a casserole dish. Cover it with olive oil, forming a layer of oil above the plant matter. Place it in the oven for two hours at 200 degrees Fahrenheit. Let it cool, then strain the oil into a jar.

USING THE HERB

Tea: Catnip Tea has a very relaxing effect on the body and mind.

Tincture: The tincture is great for sleep.

Infused oil: Store the infused oil in your refrigerator. You can use it for food purposes, internal medicinal purposes, or topical medicinal purposes.

SIDE EFFECTS AND WARNINGS

In some, it will have the opposite reaction on their gastrointestinal tract, creating stomach upsets.

For women with heavy menstruation or with pelvic inflammatory disease, catnip should be avoided. As catnip stimulates the uterus, this could intensify these conditions. And on a related note, pregnant women should avoid catnip too. The stimulation of the uterus could result in premature labor.

Catnip can cause headaches, so cease using it if you get a headache after each use. Other light conditions that could be brought about by using catnip include an increased urge to urinate or an increase in sweat–the result of its diuretic nature. You may also note that you're a bit more sleepy due to the relaxing chemical found in the plant.

An important note is that you shouldn't use catnip near times you've been scheduled for surgery. There could be interference with your central nervous system, which would make your surgery riskier. And in the same vein, it could also interfere with anesthesia, which also increases the risk of surgery complications.

FUN TIPS AND FACTS

When a cat sniffs the leaves of the plant, it experiences a hallucinogenic 'high.' This can last for 10 minutes. If the cat swallows the catnip, it acts as a sedative, making the cat sleepy and calm.

AUTHOR'S PERSONAL STORY

I've been giving catnip to my cats a few times to see if the effects are only a myth or if they're real. I mixed it into their food, and—seriously—within five minutes of eating the meal I prepared for them with catnip, all three of my cats were asleep.

13

CHICKWEED

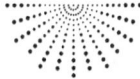

LATIN NAME

*S**tellaria*

DESCRIPTION

Chickweed is normally found as a ground crawler. It forms dense thickets the closer it gets to full maturity, with the different plants pushing up against each other so that the mass is a few inches in height. Chickweed has a single row of hairs going up each stem. This is a great way of identifying the common variety of chickweed.

Chickweed has an opposite leafing arrangement. The leaves have an ovular shape, and they taper to a point. The leaves are quite small, and while common chickweed doesn't have hairy leaves, other varieties might.

The flowers are white. Beware the difference in coloration between scarlet pimpernel and chickweed. Scarlet pimpernel has red flowers, and it has reddish splotches under its leaves. Noting this difference is important because scarlet pimpernel is poisonous and looks rather similar to chickweed. Not only that, but it sometimes grows among chickweed. Chickweed flowers are small, with hairy sepals and a hairy flower stem. It looks like there are 10 petals, but there are only 5. It looks like this because there's a deep cleft in each petal that gives it a 'v' shape. The flowers will eventually be closed up by the sepals. The sepals have hairs growing on them, so this gives the capsule that forms a hairy look. The capsule allows several small, brown seeds to grow inside.

HABITAT

It prefers damp soil and full to partial sun. It often grows on agricultural land and disturbed land.

SEASON TO GATHER THE PLANT

The plant is best gathered in January and February. At this point, there should be thick and sustainable patches to harvest from. You may already find some patches in November and December, but at that point, the chickweed will most likely still be young, and the clumps will most likely be patchy. As a note, the plant tends to die off in the summer when the weather gets too hot and then returns in the fall when things cool down once again.

PARTS OF THE PLANT TO USE AS MEDICINE

You can use its stems, leaves, and flowers.

BENEFITS AND PROPERTIES

It's highly effective for topical use. Use it for diaper rash, burns, insect bites, wounds, psoriasis, rashes, joint pain, and eczema.

You can use it to handle conditions like rabies, asthma, respiratory illness, and peptic ulcers. It's an effective blood cleanser. It can be used to reduce menstrual pain as well.

Due to its high vitamin content (including vitamin C), it handles scurvy well.

MEDICINE PREPARATION

Tea: this can be made by steeping two to three tablespoons of chickweed in boiling water for a few minutes, then straining the solids out and drinking the liquid.

Infused oil: to make this, you simply soak the chickweed in a carrier oil–choosing one that you're not allergic to and that affects your skin well. Ensure the chickweed is covered completely (you can even shake it to ensure it's properly mixed in. Leave the chickweed to soak in it for a few weeks (two at a minimum), shaking it daily so that it infuses into the carrier oil. After a few weeks, you can strain out the chickweed bits and the oil will be ready for use.

Eating it fresh

Drying out chickweed is another effective way of using it. When dried, you can add it into your food or various herbal remedies, such as poultices. You can also use dried chickweed for making tea; you don't only need to use fresh chickweed.

Powder: the powder is mainly used in teas, but it can also be mixed into other remedies. I like mixing it into honey and using a teaspoon of the mixture when experiencing one of the conditions it alleviates, such as constipation.

USING THE HERB

Tea: you can use the tea to reduce inflammation. You can also use it as a diuretic. The diuretic property of chickweed tea has led some to use it effectively for weight loss. Other benefits you could experience from this type of tea are pain relief and a calming effect.

Infused oil: you can add the infused oil to your bath or apply it directly to your skin. This is beneficial both for your skin and for your mental health.

Eating it fresh: this is great for a daily health boost. The high amount of minerals in it makes it great for healthy salads.

SIDE EFFECTS AND WARNINGS

Some individuals develop a mild rash when applying it to their skin. This isn't a common side effect, but it is a possibility. Further, some may have allergies to it, so if you've found that you're having any indications of an allergic reaction to this weed, then discontinue its use, please.

Also, if you consume it to excess, you may experience some toxicity due to the nitrate salts and saponins present in the plant. Indications of toxicity include nausea, dizziness, weakness, and cyanosis, amongst others. This won't be a problem if you consume it in moderation, and in many cases, the excess needs to be quite extreme for toxicity to become a factor.

FUN TIPS AND FACTS

In European folklore and magic, chickweed was used to promote fidelity, attract love, and maintain relationships.

Chickweed carried around was used to draw the attention of a loved one or ensure the fidelity of one's partner.

Whereas sailors used chickweed vinegar to prevent scurvy when fresh citrus wasn't available.

AUTHOR'S PERSONAL STORY

I've tried it on people with asthma and it did miracles! Within a short period, they'd have much fewer attacks. Nature is such a powerful medicine!

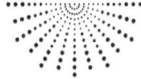

LATIN NAME

ichorium intybus

DESCRIPTION

Specimens most commonly grow between two and five feet in height. A stem will grow out of the basal leaf rosette a while into the growing season. The lower part of this stem is hairy, whereas the higher parts of the stem have no hair and black coloration. It has a long taproot, which is often harvested as a coffee replacement. When cut, this root normally exudes a bitter, milky white liquid.

The basal leaves are arranged in a rosette. On the stalk, there are a few scattered small leaves. The rosette basal leaves have toothed margins and an oblong shape. They are rather flat and tend to be between three and six inches long. The smaller leaves growing from the stem have an oval shape, are lighter green, and have a shiny surface. These smaller leaves are only about an inch long.

The flowers are usually blue or purple, but there are some pink and white varieties. The flowers normally bloom between July and October. They will open in the early morning when in blooming season, and then close once the sun gets too intense. After closing, they'll lose color and wilt. The flower generally has up to 20 petals, all tapering to a frayed tip.

The plant produces about 3,000 seeds, which get scattered across the area so that the following year's plants may grow.

HABITAT

Chicory grows—for all intents and purposes—as a weed would. You'll find that it flourishes in open fields and enjoys growing in waste areas.

SEASON TO GATHER THE PLANT

Fall is usually the best time to gather this plant. This is because most of the time, the harvesting is being done largely to get the taproot. By waiting until the plant has seeded in the fall, there won't be any sustainability issues that result. Furthermore, harvesting the taproot in the hot summer might lead to a bitter-tasting root. You'll also need to take care to not damage the root while harvesting, as it's easier to use the root when there aren't dirt-filled nicks in it.

The stem, on the other hand, is best harvested in the spring when the plant is still young and the stem isn't too hard. If this is done, it's good to only harvest part of the stand so that the rest of the stand will be able to continue its life cycle as per usual.

Similarly, also harvest the leaves of the plant when it's still young so that you don't get the bitter taste associated with more mature leaves.

The advice is the opposite for flowers. It's best to harvest the flowers close to the end of summer when they're at full maturity.

PARTS OF THE PLANT TO USE AS MEDICINE

The whole plant is used. That said, the root is normally the main attraction. You can, however, also use the stem, the leaves, and the flowers for medicinal applications.

BENEFITS AND PROPERTIES

Chicory root contains beta-carotene, which is an antioxidant that converts into vitamin A inside of the body.

Benefits that have been associated with chicory include better urine production (i.e. it's a diuretic) and a mild laxative effect that alleviates constipation. Stimulation of bile production (thus easing gallbladder conditions) is also a common effect of using this herb. You can ease liver disorders, along with conditions such as stomach upsets and a loss of appetite, by using this herb. One of the most interesting benefits that have come to light is that it has been associated with improvements in people fighting cancer.

There are also major benefits when it comes to the gastrointestinal tract. Firstly, it helps boost the production of healthy gut bacteria, which in turn helps to reduce the number of unhealthy gut bacteria present in your system. Not only this, but it has also been noted that this herb can increase mineral absorption. It doesn't only help keep the body healthy, but it can also help get the body into shape. Chicory has been noted to help reduce hormones that cause you to feel hungry, thus helping with appetite regulation. Further, a reduction in blood sugar has also been seen to result from using this herb.

As for topical use, you'll find that swelling and inflammation of the skin are reduced.

MEDICINE PREPARATION

Root: The root isn't only used for medicinal purposes. The normal use is to create a dried powder that's used as a hot drink that has a flavor similar to coffee. This is achieved most easily by using a food dehydrator and placing it on the 'nuts' setting. The chopped-up root is then left to dehydrate, and then you can crush the dried root. You can also dry it out in the oven. To achieve this, simply place the chopped-up root pieces (about an inch long each) onto an oven tray, and leave them in the oven for five to seven hours at 170 degrees Fahrenheit.

Tonic: you can make a tonic from the dried plant by steeping its leaves, flowers, and pieces of the stem in vinegar or alcohol for a month. When the month is complete, you can strain out the solid bits and use the liquid as an effective tonic.

USING THE HERB

Root: The powdered root is a good caffeine-free alternative for coffee.

Tonic: The tonic lasts long. You'll only need a few drops at a time.

SIDE EFFECTS AND WARNINGS

Some will have an allergy to this herb. If you notice allergic reactions occurring, even if mild, it's better to look for alternative herbs to handle the condition you're trying to maintain or for food purposes.

As for the increase in bile production, while this is useful for some of us, it may be a problem if you have gallstones. Consult your medical practitioner if you have or are predisposed to gallstones, and you want to use this herb.

FUN TIPS AND FACTS

Since chicory opens its flowers only between five and eleven in the morning, Carl von Linné (1707–1778) included it in the Floral Clock which he had established in the Botanical Garden at Uppsala.

AUTHOR'S PERSONAL STORY

When I stopped drinking coffee because it was affecting my well-being, I replaced it with a cup of chicory 'coffee.' I love brewing my chicory coffee every morning and enjoying it with an audiobook. It prevents me from having any digestive issues all day long.

15

CLEAVERS

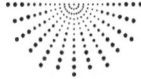

LATIN NAME

Galium aparine

DESCRIPTION

The plant grows to approximately five feet in height. It is also rather large and bushy.

The leaves of this plant are arranged in whorls on the end of stems. Each whorl has eight leaflets, each of which is lanceolate. They have prickles on them that give your hand a sticky feeling when you come into contact with the leaf. The scent you get from the plant once you've cut it and left it to dry is that of freshly-mown hay.

The flowers are very small, and they're shaped like small stars and tend to be white or greenish. They bloom in spring and summer, growing in clusters of two to three flowers per flower head. The stem bearing the flower cluster will come out of the axis of the leaf.

The fruit takes the form of hooked burrs that grow in clusters (normally up to three in a cluster). They tend to hook onto people and animals as a form of dispersing for propagation.

HABITAT

Wet areas are generally its preferred location. They grow all over the world, and the Northeastern United States is no exception.

SEASON TO GATHER THE PLANT

I would suggest that you gather the plant in May and June as it's flowering. This is the best time, in my opinion, as the flowers are great for medicinal purposes.

PARTS OF THE PLANT TO USE AS MEDICINE

The leaves and flowers of cleavers are generally the best parts to use for herbal medicine.

BENEFITS AND PROPERTIES

There's a long list of benefits that you could get from using cleavers. This includes benefits from topical application. When applied topically, you can alleviate a range of skin conditions, such as eczema. It's also a great inflammation reducer. A further topical use that you could benefit from is wound care. Using it on wounds can help to speed the wound's recovery and can reduce inflammation around the wound.

There are benefits to the body's excretory system. It helps ease constipation and can increase urine output. It also increases sweat production, which goes hand in hand with its detoxification properties.

There is a special benefit for women. This herb helps stimulate the uterus, which can in turn lead to less intense menstrual cramps. Just don't use it while you're pregnant because you don't want to stimulate your uterus and accidentally cause premature labor.

There are also a few specific conditions that can be reduced by using cleavers. This includes tonsillitis, fevers, swollen lymph nodes, jaundice, and body spasms.

One of the most enjoyable benefits for me on a personal level is that it helps boost my energy levels. This is great for when you're starting to feel lethargic or burned out.

MEDICINE PREPARATION

Tea: The tea is best made by adding two to three teaspoons of the wet or dried herb to a cup of boiling water, and allowing it to steep for 10 to 15 minutes. Once steeped, strain it and enjoy.

Tinctures: Soak a handful of the fresh cleaver in alcohol (I normally use vodka) over a few nights. Strain out the plant mass. You can then take a small amount with an eyedropper under your tongue when needed.

Pulp: Apply the pulp directly to bug bites or wounds for instant relief.

Wash: Make a very strong cup of tea with either fresh or dried cleaver. Then strain out the pulp and apply the wash to skin that's not feeling well (e.g. itchy or inflamed).

USING THE HERB

Tea: The tea can be a good energy booster.

Tincture: The tincture can last for months, or sometimes even years. Just remember to keep it in a dark bottle.

Pulp: The pulp should only be used once, otherwise things could become unhygienic.

Wash: You can use it on wounds for a quick and relieving clean.

SIDE EFFECTS AND WARNINGS

The only real side effect I can mention here is an allergic reaction. Other than that, there are no known side effects of note.

FUN TIPS AND FACTS

It's possible to freeze cleaver juice for a few months. This way you'll be able to have a juice stock for poultices etc. even when the rest of your cleavers plant stock has dried out.

AUTHOR'S PERSONAL STORY

Since ancient times, this plant has been used as a diuretic.

When I help people detox, I add this plant sometimes, and after a month the person comes back to me and looks changed—slimmer, more energized, and overall healthier.

16

COMFREY

LATIN NAME

Symphytum officinale

DESCRIPTION

Comfrey grows between two and five feet high, and it reaches about two feet in width.

The roots are black in appearance, and when you break one open, you'll see that the inside is white and exudes juices.

The leaves have an alternate arrangement closer to the base of the plant and then transition to an opposite arrangement higher up on the plant. The leaves are larger closer to the base and get smaller higher up on the plant. The leaves closest to the base can be up to eight inches long. The leaves at the base also have a slightly different shape than those higher up on the plant. Lower down, the leaves have a broad base and taper to a tip, whereas higher up, the leaves tend to be broad along the whole margin (i.e. the leaves are oblong) and only narrow near the ends (base and tip). The leaves have a hairy texture above and below, but they are slightly more hairy on the underside. These stiff hairs give the leaves a rough texture. The veins on the leaves have a net arrangement, and the margins are smooth.

The flowers can vary quite a bit in color. The most common varieties are white, cream, and purple. But you can get other colors, and you can also get striped varieties with more than one color. The flowers grow in rather dense clusters. Individual flowers are small and have a bell shape to them. The various comfrey species can cross-pollinate with each other, which is one of the reasons there's such a diversity of color arrangements.

Its fruits are smooth on the surface and angle to a concave base. They are small and have a brown-black color.

HABITAT

Comfrey prefers to be in damp and shady spaces. This makes them most common near rivers and streams. You'll also be able to find them in woods and meadows.

SEASON TO GATHER THE PLANT

You can gather the plant multiple times a year because it's hardy and regrows even when you trim it within two inches of the ground. The first harvest is possible at the start of spring. After that, you'll be able to re-harvest the same plant approximately every month and a half. It's suggested that you only gather the leaves. You can gather the roots too, but the leaves aren't as strong, making them a better choice.

PARTS OF THE PLANT TO USE AS MEDICINE

The leaves and roots can both be used. It is, however, better to use the leaves as the roots contain a higher concentration of a chemical that can have toxic effects on the human body.

BENEFITS AND PROPERTIES

This herb can be used to improve ulcers, coughs, diarrhea, bloody urine, heavy menstruation, hemorrhoids, gum disease, and a sore throat. It's best to not overuse it for these internal applications. So if you use it for internal applications by drinking the tea, make the tea mild and don't use it often.

Topical application is where this herb shines. It works particularly well for reducing pain. This applies to pain in the joints, such as with arthritis and gout, as well as pain in the muscles. It's also effective at improving blood circulation, healing wounds (use on closed wounds, not open wounds), and repairing bruises.

MEDICINE PREPARATION

Tea: Make a tea with the herb by steeping it in warm or boiling water for a few minutes. Don't use much of the herb in one cup–I find that a teaspoon's worth of shredded leaves is sufficiently effective.

Ointment/cream/poultices/embrocations: I suggest drying out the leaves and then crushing them into a fine powder first. Once this is done, adding a small amount of the powder to a fatty substance (such as lanolin or coconut oil) and mixing it in evenly will do the trick. [Note: embrocation refers to a pain-relieving rub.]

USING THE HERB

Tea: Don't take a lot of tea. It can become toxic when used internally.

Ointment: This herb is especially effective for external applications. Pharmaceutical companies even use it as an ingredient in healing creams. I find that it's particularly effective when applied externally to painful parts of the skin and muscles. That said, only use it when needed as everyday use can cause toxicity.

SIDE EFFECTS AND WARNINGS

Comfrey contains pyrrolizidine alkaloids, which have a very strong effect on the human body. Negative effects that can result from these alkaloids include liver damage, cancer, and lung damage. The Food and Drug Administration has suggested that products containing comfrey that are used for oral consumption be taken off the market.

The root contains an especially high concentration of these chemicals, making it preferable to use the leaves (as they have a lower concentration).

You should also avoid using it on open sores or broken skin, as well as avoid using it for protracted periods.

FUN TIPS AND FACTS

If you plan on growing this herb in your homestead or garden, you'll see that the plants around it tend to do quite well. This is because comfrey encourages the presence of earthworms in the soil surrounding it.

The Ancient Greeks were already onto this use of the herb. 'Comfrey' comes from an Ancient Greek word meaning "to make grow together or bind."

AUTHOR'S PERSONAL STORY

There was this one gentleman who came to me for help recovering from bumps and bruises he got while playing football. I gave him a comfrey poultice to use on his bruises and sore muscles, and he never visited me again. When I bumped into him a few years later, he said that it helped so much that he's had a small container of comfrey poultice on standby for bumps and bruises ever since.

17

COMMON MALLOW

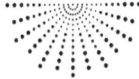

LATIN NAME

Malva sylvestris

DESCRIPTION

The plant can get up to about three feet tall. It is, however, a creeper. This means that it might not reach its full height, but rather, it spreads along the ground. The stem on this species of mallow has fine hairs on its surface. Common mallow stems branch in all directions, like any crawler.

The leaves of common mallow are alternate in their arrangement. They have palmate leaves that are approximately two and a half inches long and three inches wide. The leaves have a circular, kidney-like shape, with five or more shallow lobes. The leaves' margins are toothed, but not very deep, giving each tooth a rounded appearance. These leaves are attached to very long petioles (twice as long as the leaf itself). The leaves are a bright green color and have very fine hairs along their surface. Note that the leaves also have a deep indent at the base, thus they don't have a flat appearance.

The flowers are violet or whitish. The white flowers sometimes have light violet lines running along their length. There is a unique musky scent to the common mallow, and you'll find some perfumes that try to imitate this scent. Flowers on common mallow grow at the end of the peduncle, normally having between one and three flowers per peduncle. They are rather small—about three-quarters of an inch in width. There are five petals on each flower, with each petal having a small groove or notch at the tip. There are five calyxes below the petals, each shaped like an egg or oval at the tip. There is one female part (pistil) in each flower and a group of male parts (stamen) inside the flower.

The fruit of common mallow looks like a wheel with a tire on it. It is small and contains flat seeds, with the outer layer of seeds sometimes being hairy.

HABITAT

You can find them in sunny areas that are mildly moist to slightly dry. They enjoy areas that have been disturbed and flourish at establishing themselves in these areas. You'll find that they commonly occur in fields and open areas such as farm lots.

SEASON TO GATHER THE PLANT

The flowers generally bloom in the summer, although some bloom in the spring or autumn. You can harvest these flowers anytime you notice them in bloom.

As for the leaves, they are best gathered in spring when the leaves are still young. If you're planning on gathering leaves after the spring, stick to harvesting softer, younger-looking leaves.

The roots are best harvested over the fall period. This is because the plant is dormant during the fall.

PARTS OF THE PLANT TO USE AS MEDICINE

You can use the roots, the leaves, and the flowers. All three are practical when used medically.

BENEFITS AND PROPERTIES

The best benefits are those you'll get from topical application. This includes skin soothing, wound healing, and inflammation reduction. The benefits you may expect from internal use are diarrhea relief, bladder problem relief, and upset stomach relief. The mouth and respiratory tract can also benefit from the use of this plant. When using it for this purpose, you'll be able to treat conditions such as throat irritations, dry coughing, colds, and bronchitis.

MEDICINE PREPARATION

Tea: You can use the flowers, the roots, or the leaves for a good cup of mallow tea. Place three to four teaspoons of dried mallow into a cup of boiling water. After steeping it for 10 minutes, you can remove the dried mallow bits. For internal use, you would drink it at this point. For external use, you would soak a cloth in it and apply the cloth to an ailing area on your body surface, or you could splash some of the tea on directly.

Pulp: the best pulp is obtained by crushing fresh mallow bits together and applying them to the afflicted area of your body. You could also mix it in with natural fats and oils, but I tend to apply the pulp directly.

Cold infusion: a cold infusion just might be the most effective method of using mallow, both internally and externally. Take a decent amount of dried mallow (approximately 2 tablespoons) and steep it in a quart of water overnight. You will then be able to drink from it for internal benefits or soak a cloth in it and apply it to yourself for external use.

USING THE HERB

Tea: This tea can be used for both internal and external use.

Pulp: Placing the pulp in a rag and placing it over the afflicted area can also do the trick. In this case, you would be using it as a compress.

Cold fusion: This is the most effective preparation, so I suggest it. It's particularly effective when you're working through a cold and fever.

SIDE EFFECTS AND WARNINGS

The main side effects are nausea, diarrhea, and indigestion. That said, this is normally linked to an allergy, so if you experience any of these three while using it, you may find that you're allergic to the plant.

FUN TIPS AND FACTS

Mallow has been used by the Iroquois as a love medicine. It's eaten and then vomited up as a ritual act.

AUTHOR'S PERSONAL STORY

As a kid, I used to hike in the mountains a lot with my parents. Whenever I got cut, if mallow was around, my parents would put it directly on the wound, and it promoted my healing.

18

CRAMP BARK

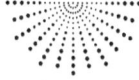

LATIN NAME

Viburnum opulus

DESCRIPTION

Cramp bark is a large (13 to 15-foot-tall) shrub. Its leaves are three-lobed and between two and four inches long and wide. They have a wrinkled surface with their venation impressed quite sharply. The margins are rough and jagged, and the leaves are green in budding and mature stages then transition to a purple color in the fall.

The flowers come in a flat-headed cluster, and each floret has five white petals. The flowers on the outer part of the head are larger (approximately half an inch in diameter), and the flowers in the center are smaller (about a fifth of an inch in diameter). Only the central flowers are fertile. The plant is a monocot, so there aren't separate male and female shrubs.

The fruit is a small, translucent red drupe with a diameter of approximately a third of an inch. The drupes grow in clusters that hang down. There is only one seed in each drupe.

The bark is brown and has a slightly cracked texture, and sometimes you'll find brown warts on its surface.

The shrub has a slightly bitter scent to it–especially the bark. The taste is also bitter, including the taste of the fruit.

HABITAT

Most cramp bark shrubs are found in nutrient-rich, moist soil and exposed to direct sun or partly in the shade. This means you'll mostly find them at the edge of forests, in woodlands, or in scrublands.

SEASON TO GATHER THE PLANT

The bark is best gathered in early spring or early fall before the leaves have started changing color to purple.

Flowers can be harvested in late spring and early summer.

Fruit is ready for harvesting in the fall.

PARTS OF THE PLANT TO USE AS MEDICINE

You can use the roots, bark, leaves, and flowers for medicine, but the bark and fruit are most commonly used.

BENEFITS AND PROPERTIES

Cramp bark is one of the most beneficial herbal medicines I know of. It handles and treats a host of conditions. It's known mostly for treating conditions related to muscle pain and muscle spasms. Rheumatism, heart palpitations, painful menstruation, bladder spasms, colic, and overly strained eyes can all be treated with cramp bark. One of the more common problems business people face is tension headaches, and these, too, can be alleviated with cramp bark.

Breathing conditions can be alleviated with it too. If you have asthma or difficulties breathing, I would suggest cramp bark.

You can reduce the effects of heart disease with it, as well as circulatory issues and high blood pressure. Just don't mix it with other blood pressure medication if you're already taking such.

A few other things that can be remedied using cramp bark include swollen glands and mumps, fluid retention, urinary conditions and kidney problems, irritable bowel syndrome, and some emotional issues.

Another condition that many people have issues with is difficulty sleeping. Cramp bark has a sedative effect, which means you can use it to finally get those elusive hours of sleep you've been needing. As this is one of the effects of this herb, avoid using it when you're tired and trying to stay awake (such as when taking exams after staying up to study).

MEDICINE PREPARATION

Decoction: Put a cup of water and two teaspoons of dried bark in a pot or pan. First, you'll need to bring the water to a boil. After this, you should bring it down to a simmer. Keep it simmering for 15 minutes or so, and then take it off the heat. Let it cool down, at which point it will be ready to use.

Tincture: You'll need a mason jar, a cup of water (preferably distilled), a cup and a half of very strong grain alcohol, and 4 ounces of dried cramp bark. Place all the ingredients in a mason jar, seal it, and shake it. Leave it to stand for a month to a month and a half, giving it a shake every few days. Once you've left it long enough, you can strain out the bark pieces and keep the tincture in a container that seals well.

USING THE HERB

Decoction: You can have up to three cups of decoction per day. The decoction is really helpful for painful menstrual cramps.

Tincture: You can have up to two teaspoons per day. The tincture is especially useful for muscle cramp relief of any kind (from your heart to your calves).

SIDE EFFECTS AND WARNINGS

There aren't many side effects with cramp bark. You may find that the berries can have a toxic effect if too many are eaten. But for medicinal purposes, I generally suggest using the bark rather than the berries because of its high degree of effectiveness.

You may also want to consult your doctor if you're on blood pressure medication, and you're planning on taking cramp bark. Some medications interact with cramp bark and can cause unintended side effects.

FUN TIPS AND FACTS

It has been used for millennia to treat cramps! In fact, the Meskwaki tribe of Wisconsin used it mainly to treat women's menstrual cramps.

It has also been used by Native American tribes to take the place of tobacco for smoking.

AUTHOR'S PERSONAL STORY

The plant name says what it treats: it helps for cramping muscles and it's a true elixir for that! I've used it many times over the years for cramp relief after a long day of work.

19

CRANBERRY

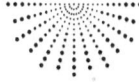

LATIN NAME

V accinium macrocarpon

DESCRIPTION

Cranberries are bushes with thin, creeping stems. They reach between two to eight inches in height but stretch out for about seven feet in length and are green and brown.

Their leaves are arranged in an alternate pattern and are rather small (approximately half an inch long). The evergreen leaves have an oval shape and some have a slightly shiny surface.

They produce small pink flowers with four petals. You'll see the stamen protruding from the flower, and in some species, there will also be a purplish color in the center of the flower.

. . .

THE BERRIES THEMSELVES ARE WHITE WHEN THEY START growing, turning a bright red color as they mature. There are often spots on the berries. The average berry will be about the size of a raisin.

HABITAT

Cranberries love the wet, boggy ground. They are very finicky when it comes to where they grow. You'll find them in colder areas that are open and in direct sunlight. They have an acidic soil preference.

You're most likely to find them in Massachusetts, New Jersey, Oregon, Washington, and Wisconsin.

SEASON TO GATHER THE PLANT

You can harvest the leaves year-round since it's evergreen. The flowers should be available for harvesting in June and July, and the berries are normally ripe for picking during September and October.

PARTS OF THE PLANT TO USE AS MEDICINE

The leaves and berries are most useful for medicinal purposes. Generally, the other parts aren't used for medicine or food.

BENEFITS AND PROPERTIES

The most well-known benefit of cranberry consumption and taking cranberry medicines is that it combats urinary tract infections.

Your oral health can likewise be improved by taking cranberries. The plaque and bacteria that like to hang around in your mouth find it difficult to cling to your teeth and gums after consuming cranberries.

Heart-related problems can be reduced by taking cranberries. This includes cardiovascular disease and blood disorders.

If you have a sore stomach, then cranberries make a good solution. Cranberries can also assist with a bad appetite and in reducing your body mass index because it's a good blood sugar regulator.

Something that may give many people hope is that cranberries have been linked to cancer healing. There is far too little testing to indicate a definitive verdict, but thus far, the tests are promising.

MEDICINE PREPARATION

Fresh and dried berries: Probably the easiest and most direct way to get the benefits is to consume cranberry fruit directly. They're tasty to snack on and can also be used to make tasty salads. Fresh cranberry leaves can also make a nice addition to any salad.

Juice: I prefer to cold press the juice for a pure result. Diluting with water is a good idea because it can be quite strong.

Pulp: the fresh leaves make a great pulp. Simply crush them and apply them to the afflicted area.

USING THE HERB

Berries: Eating the berries is the best way to handle oral health issues using cranberries.

Juice: Juice is the best remedy when it comes to urinary tract infections.

Pulp: This will help with bruises, wounds, burns, and sprains.

SIDE EFFECTS AND WARNINGS

The side effects associated with cranberries are mainly runny stool or upset stomach. You could also find that you bleed more easily. Also, if you're predisposed to kidney stones, a high dose of cranberries can result in developing kidney stones.

Additionally, the drug warfarin can interact with and produce unintended side effects if you have large amounts of cranberries.

FUN TIPS AND FACTS

Cranberries are native to North America, soo if you like buying local food and goods, this is about as local as it gets.

AUTHOR'S PERSONAL STORY

As mentioned in the other book, I've helped many women with bladder infections. Homemade cranberry juice does pure magic to people who suffer from frequent bladder infections.

20

DANDELION

LATIN NAME

Taraxacum offinale

DESCRIPTION

These plants have a long and thick taproot topped by a cluster of floral leaves. The plant can reach a size of 25 inches high and 20 inches across.

These leaves are long and lobed. If snapped, the leaves produce a white, milky substance called latex.

Rising from the basal leaves are the flower stems. These are hollow and don't produce any branches. They are crowned by a bright yellow flower head made up of a large number of florets.

Eventually, this flower head is replaced by an ovary that grows a bunch of seeds. The seeds produce a white puffy ball, and if you blow on the ovary, several of the seeds will blow off.

If you smell it, the flower gives off a relatively sweet scent and the leaves have a bitter, citrusy scent.

HABITAT

The optimum habitat for a dandelion plant is a cool, shaded meadow. That said, they can also thrive in hot, direct sunlight. You may commonly find them in all sorts of pastures and disturbed areas too.

SEASON TO GATHER THE PLANT

The plant is most likely to be ready for harvesting in the fall and the spring.

PARTS OF THE PLANT TO USE AS MEDICINE

You can use the whole plant for medicinal purposes. This includes the flower, the stem, the leaves, and the root.

BENEFITS AND PROPERTIES

Dandelion is an especially good diuretic. Use of it will result in better regulation of urination.

It's commonly also used as an appetite stimulant, so if you haven't been able to stomach anything for a while, try out some dandelion as a solution. It's also good at increasing the effectiveness of your digestion, resulting in better metabolism of carbohydrates and reduction of fat absorption. Not only this, but your blood sugar regulation can also be improved using dandelion. This combination of benefits can help with weight loss.

Your immune system will benefit too. The plant is a good antiviral and antibacterial. Hepatitis B, specifically, has been shown to slow in its spread when one takes dandelion.

Some of the organs in your body that will especially benefit are your liver and gallbladder. This plant can be an effective means of detoxifying both.

There are skin benefits to using dandelions as well. In addition to its ability to reverse ultraviolet damage, it's an effective antioxidant, resulting in less cell damage from free radicals. Further, inflammation is effectively reduced by using this herb.

Cancer studies have found that dandelion can be of benefit in treating colon, liver, and pancreatic cancer. While more research needs to be done regarding this, there have been cases in which it has yielded results.

In terms of properties, dandelions are full of natural vitamins and minerals. You'll find that it's high in vitamins A, C, E, and K. It's also high in calcium, folate, iron, and potassium.

MEDICINE PREPARATION

Fresh salad: Add some fresh basil leaves along with dandelion stems and flowers to your salad.

Tea: You can use dry or fresh dandelion to make the tea. If you use the flowers or leaves, you need only steep it for a few minutes. If you're using dried root, put two tablespoons and a cup of water in a saucepan. Then bring it to a boil for three minutes before switching it off. Leave it to steep in the saucepan for another half hour, at which point the tea will be ready to drink.

Salve: You'll first need to make infused oil before getting started on the salve. Take about two cups full of fresh dandelion and put it in a jar. Cover it with a carrier oil of your choice (olive oil is quite good for this purpose). Leave it to soak for about two weeks, at which point you can strain out the dandelion pieces. At this point, heat the infused oil and melt in a few tablespoons of beeswax. After the beeswax, melt in a few tablespoons of shea butter. Stir until everything is incorporated quite well, and then leave it to cool off. At this point, you can place it in jars for your daily use.

USING THE HERB

Fresh salad: The flowers have a honey-like flavor, while the leaves have a slightly bitter but earthy flavor. Dandelion flowers and leaves can be consumed in the same quantity and frequency as any other salad leaves you might find in the supermarket. It's easy to incorporate into your daily salad routine.

Tea: Using the flowers gives a nice sweet flavor to the tea, while using the root gives you an earthy flavor similar to chicory. Adding a bit of honey to tea made from the root makes an extra tasty beverage.

Salve: The salve can be applied to any sun-damaged areas and any inflamed areas. It's great for dry hands too.

SIDE EFFECTS AND WARNINGS

The side effects include skin irritation and mouth sores if you're allergic to dandelion or latex.

Due to its acidic nature, it can increase stomach acid and contribute to heartburn.

If you have chronic gallbladder problems, consult your doctor before using it.

FUN TIPS AND FACTS

Many people see dandelion as weed and throw it in their compost! Instead of just being a nuisance, this plant grows in spring for people to detox their bodies from the heavy foods they've been eating throughout the colder seasons!

AUTHOR'S PERSONAL STORY

In my home, we eat fresh dandelion leaves, stems, and flowers every day for a month. Usually, we make a salad, and it's either the main dish with some hard-boiled eggs or a side dish.

21

ECHINACEA

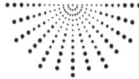

LATIN NAME

Echinacea purpurea

DESCRIPTION

Echinacea is a hardy plant that comes in an upright, rounded shape. The stem is stiff and covered with hairs and usually has purple stripes running down its length. The plant grows to about three and a half feet at a maximum height and about two feet wide.

Simple leaves protrude from the branches, coming out in an alternating pattern. The leaves have a serrated margin, and their color can be various shades of green. The leaves have a lanceolate shape–going from a broader base to a tapering tip. Typically, they grow three to six inches long and one to three inches wide. The leaves gradually get smaller as they approach the top of the plant. At the bottom of the plant, the leaves have stalks, but as they get closer to the top of the plant, they become stalkless—note that the stalks are green or brown. Each leaf has between three and five clear veins. And as for the surface of the leaves, you'll find they can be either smooth or hairy.

Echinacea has a composite flower head, and the outer petals of this plant are usually pink, pink-purple, or purple. These petals range between 15 and 20 in number. Each petal grows from one and a half to three inches long and has an approximate width of a quarter inch. The tips of the petals typically have three notches. The center of the flower head consists of an orange-brown disc or cone with a mass of small yellow flowers on its surface.

The fruit of the plant consists of a cone with sharp spines. This cone is a seed head, and it has a purple or brown color.

HABITAT

You'll find these plants in soil that ranges from moist to slightly dry, usually easily draining soil. The plant prefers either direct sunlight or semi-shade. Echinacea can commonly be found in the East, Southeast, and Midwest.

SEASON TO GATHER THE PLANT

You should harvest echinacea in its second year for best results. It's not always possible to know how old the plant is in the wild, but you can generally tell by its size if it's fully matured or not. The flowers mainly bloom during the summer–which is the right season to harvest them. The seed cones appear in the fall, and you should usually wait until they've dried and gone brown before harvesting them. You can forage for the roots and the leaves at any time. That said, it's best to harvest the leaves in the summer, and you should ideally harvest the roots in the fall.

PARTS OF THE PLANT TO USE AS MEDICINE

The leaves and roots are the most used parts of this plant. You can also use flowers and stems, especially when preparing tea. You can use the seeds, but this isn't a common practice.

BENEFITS AND PROPERTIES

The benefits of using echinacea include improvement of skin conditions. It can help hydrate dry skin, and you can use it to assist with wrinkly skin. You can also use it to alleviate acne. Athlete's foot and inflammation can also be treated topically with echinacea

Echinacea is useful for tackling infections due to its antiviral properties. It can be used to remedy urinary tract infections, ear infections, cold sores, colds, and the flu. If you have nasal issues, echinacea can also help with that. Hay fever and sinusitis can both be relieved with echinacea.

Other conditions that can be alleviated or eradicated using this plant are pain, high blood sugar, anxiety, and wounds that won't heal.

MEDICINE PREPARATION

Juice: Wash the plant then simply squeeze the fresh plant matter so that the juice comes out. Add a teaspoon of the juice to a glass of water and drink it.

Tea: Dried and crushed leaves, roots, and flowers are best for the tea although you can also use the fresh versions if you want. Boil water and then let it cool down for a minute. Pour the hot water over the crushed echinacea and let it steep for 15 minutes. Strain it and drink.

Tincture: This requires vodka and two whole echinacea

plants. Clean off the plants and place them in a jar. Cover the plants with vodka, then seal the jar. Leave it for one to three months in a cool place, shaking it every few days. You can then strain the liquid into a container for use.

Salve: you'll need three and a half ounces of carrier oil, half an ounce of beeswax, and one to four handfuls of dried leaves and flowers. You'll start by making infused oil. To do this, clean the echinacea plants and put them in a jar. Then cover the dried plant bits with the oil and seal off the jar. Leave this out of direct sunlight for a month to a month and a half. At this point, the oil should be strained into a pot. You then melt the beeswax into the infused oil and pour it into a container or containers so that it congeals. This is now ready for use.

USING THE HERB

Juice: The juice can also be used in skin pastes for wound care or skin conditions.

Tea: Tea can be taken daily for five days when you have a cold or flu to help get rid of the condition.

Tincture: Use about 5 drops of the tincture at a time. I prefer dropping it into a glass of water because it tastes very strong. The shelf life is very long—it lasts years.

Salve: Use this for skin conditions or wounds. The shelf life is between nine months and a year.

SIDE EFFECTS AND WARNINGS

Side effects that can commonly occur are feeling nauseous or dizzy, having a sore stomach, or having dry eyes. If you take it orally, such as with the tea, juice, or tincture, then you may experience a numb tongue for a while.

People who are allergic to this herb experience effects

such as itchy skin, hives, nausea, shortness of breath, and pain in the stomach.

If you have an autoimmune disease, you should avoid this plant because it can affect your immune system. This is especially the case if you're on drugs that affect the immune system because you may experience an immune boost and drug interactions. Liver medication may also interact with echinacea, so consult your healthcare advisor before using it if you're on such medication.

FUN TIPS AND FACTS

This plant symbolizes strength and healing because of its properties.

AUTHOR'S PERSONAL STORY

A patient who would come to me and tell me that every autumn they get the flu stopped coming because echinacea helped them a lot with flu prevention by strengthening their immune system.

22

ELDER

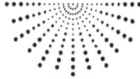

LATIN NAME

*S*ambucus

DESCRIPTION

This is a large shrub (10 feet tall in some cases) with a short gray trunk that has smooth bark. The bark is soft, with corky bumps and furrows. When the shrub grows older, the bark becomes rougher and gets a brown-gray tinge. This older bark is marked by small fissures. The twigs are yellow-green and porous, and the pith inside the branches is white and soft.

The leaves of this shrub have an opposite arrangement. These leaves are pinnate, having between three and 11 dark green leaflets–although between five and nine is the most common number. The margins on these leaves are serrated. The leaf holding these leaflets has a total length between two and 12 inches.

The flowers are small, white, and have five petals and five stamens each. They form a clustered flower head that ranges between six and 12 inches in diameter. The flowers have a lovely floral, creamy scent.

The fruits (a type of berry) are very small–just over a tenth of an inch in diameter. They hang down in dark purple, drooping clusters.

HABITAT

Elder prefers direct sun and soil that ranges from moist to slightly dry. It grows well in the wild areas of the US–specifically to the east of the Rockies.

SEASON TO GATHER THE PLANT

If you do decide to harvest the leaves, it is best to harvest during April. This period before the flowers grow is best, but you can harvest the leaves until fall.

The flowers are best gathered in May and June. When gathering them, preserving the pollen by using a container or pot is ideal. The pollen has good qualities and adds to the flavor.

The berries should be harvested close to the end of summer. At this time they are soft, and their color is dark purple.

PARTS OF THE PLANT TO USE AS MEDICINE

The only parts you should use internally are the flowers and the berries. The rest can be toxic to your body. The leaves can be used but should only be used externally to alleviate bruises or pain. Please be cautious if you decide to use the leaves.

BENEFITS AND PROPERTIES

On a topical level, elder normally only works on mild skin conditions, joint pain, muscle pain, and inflammation. However, the plant is a winner when it comes to oral usage. It can be used for both mental and physical conditions.

In terms of mental conditions, it is effective at dealing with stress. I've used it for this purpose, and I know many other people that have used it for this purpose too. In terms of the brain itself, elder can be used to keep epilepsy in check, and it can assist in relieving headaches.

On a physical level, you can use it for immune-system-related issues and for boosting certain organs. Immune system issues that can be cured or alleviated using this plant are flu, fevers, and colds. Breathing difficulties due to infections can be improved with the elder herb as well. HIV and AIDS can also be more easily managed with this herb. In addition, using elder can help your heart and kidneys.

On another note, elderflowers can be used for alleviating allergies in many people. It's especially good for treating sinus issues due to allergies.

MEDICINE PREPARATION

Infusion: The infusion can be made hot or cold. A hot infusion is made by putting one to three tablespoons of the flowers (dried or fresh) into a cup of water. Bring this water to a boil then switch off the heat and let it steep for 15 minutes. The cold version is made by putting a few tablespoons of the flowers into a jar of water then leaving it to infuse overnight.

Berries: The berries should be cooked because when raw, they can cause nausea. Throw them into any dish that could use a slightly sweet and tart flavor.

Cough syrup: You'll need a cup of fresh, dark elderberries, a cup and a half of raw honey, and four cups of distilled water. Mash the berries into the water in a saucepan. Bring the water to a brief boil then down to a simmer for 45 minutes. At this point, the mixture will have reduced. Switch it off and let it cool, adding the honey once it's finished cooling.

USING THE HERB

Infusion: The infusion can be diluted into a wash and used on the face. This is an effective agent for evening out your skin tone.

Berries: Due to their high vitamin content, adding the berries to your food is good for maintaining overall health and wellness.

Cough syrup: The cough syrup should last for about three weeks. It can be frozen too, in which case, it should last for several months. Just let the frozen cough syrup thaw when you want to use it.

SIDE EFFECTS AND WARNINGS

The plant contains a chemical that can be used to make cyanide. This chemical can induce nausea, diarrhea, and vomiting. The flowers and berries are particularly liable to induce these reactions.

If you're using medication for diabetes, then avoid using elder. It lowers blood sugar, and this may cause interactions and unintended side effects.

FUN TIPS AND FACTS

In some Christian mythology, people believed that burning elder wood would aggravate the Devil and bring death because it's a sacred plant.

AUTHOR'S PERSONAL STORY

I've used this lot with people who have had fevers. The tea makes people sweat a lot, and drinking it before bed with a spoon of honey would heal them overnight.

23

FEVERFEW

LATIN NAME

Tanacetum parthenium

DESCRIPTION

The stem of this plant is thin and circular and has a whitish fuzz closer to the top of the plant. The stem is usually a light green color, and the plant grows approximately two feet high.

The leaves are arranged alternately and are green or yellow-green. Their margins are bluntly toothed, with the teeth being rather large. In terms of size, the leaves range from one and a quarter inches long to six inches long, with the smaller ones closer to the top of the plant. The width of the leaves is between half an inch and two and a half inches. Feverfew leaves are pinnate, made up of three or five leaflets. The shape of the leaves is between an oval and a triangular oval. The leaflets making up the leaves are lanceolate or ovate. The upper part of each leaflet is smooth, but the lower part has a soft feel due to hair-like secretions from glands in the leaf. The leaves have petioles, with the ones close to the bottom being up to two inches long and the ones on the upper parts of the plant being very short to nonexistent.

The flowers on feverfew have white petals around a central disk of yellow florets. There are approximately 12 petals on a healthy flower, with each petal having a few teeth at the tip. The disk in the center sometimes looks quite puffed up with all the yellow florets in bloom. You'll find that the flowers give a relatively strong aromatic smell.

The fruit occurs in the form of achenes that are ribbed and brown. They are small and occur in clusters on the head where the flower once was.

HABITAT

Feverfew prefers disturbed landscapes. These and other wild areas such as fields are the most common locations for the plant. It prefers sunlight and well-draining soil.

SEASON TO GATHER THE PLANT

The flowers bloom during the summer, which is the best time to gather them. They can sometimes bloom until the start of fall. The leaves can be harvested the whole year round, but it's best to leave them during spring so the plant can sustain itself while growing flowers.

PARTS OF THE PLANT TO USE AS MEDICINE

The leaves are the main part used for medicinal purposes. You can, however, use the whole plant.

BENEFITS AND PROPERTIES

The main use of feverfew is as a pain remedy. This is especially so when it comes to headaches and migraines. Some properties of feverfew reduce the dilation of blood vessels in the brain, thus reducing pain in your head. Migraines won't be fully cured, but they can be managed and in many cases, their occurrence will be drastically reduced (if the migraines are chronic). Side effects of migraines, such as vomiting and nausea, should also be alleviated when using feverfew.

Inflammation is also effectively remedied using feverfew. Inflammation due to arthritis is particularly prone to relief using this herb. Inflammation in the colon (colitis) can be greatly eased by ingesting the herb as well. Insect bites and their related inflammation can be soothed by using it topically.

In addition, feverfew can ease menstrual symptoms. Additionally, if a period is delayed, you could use it to start your cycle, thereby making periods more regular. Further, when a woman is having strong contractions while giving birth, feverfew should help reduce the intensity of the sensations being experienced.

The digestive system can also benefit from taking feverfew. When your stomach is feeling funny, the tea can help.

MEDICINE PREPARATION

Leaf: Simply eat one or two fresh leaves.

Tea: Pour a cup of boiling water over a teaspoon of dried feverfew leaves. Let it steep for a few minutes, then strain and drink.

Compress: Pour a cup of boiling water over four teaspoons of dried leaves. Let it steep for 20 minutes or more. Soak a cloth in the liquid and apply it to the affected area.

USING THE HERB

Leaf: This is very effective for headaches. Don't eat more than one or two at a time because you could get sores in your mouth.

Tea: Use the tea when you have a headache. It's effective but won't leave you with mouth sores like eating the leaves directly might.

Compress: The compress is effective for recovering from bruises.

SIDE EFFECTS AND WARNINGS

Mouth ulcers are the most common side effect. This occurs when taking the fresh leaves orally.

Withdrawal symptoms can take place when you cease using it after a long period. Sleeping difficulties, sore muscles, anxiousness, and headaches are all symptoms you can expect when going through feverfew withdrawal.

Some have allergies to feverfew, so test it out before using it at full dosage levels.

During pregnancy, you shouldn't use feverfew. You can use it while going through labor, but cease using it when you begin breastfeeding.

Interactions can occur with blood pressure medication, especially if you're on blood thinners. Cytochrome P450 3A4, a type of protein used in many drugs to assist with metabolizing the drug in question, can also interact with feverfew. Due to the variety of drugs that contain this type of protein, it's important to consult your healthcare professional before planning on using feverfew.

FUN TIPS AND FACTS

In Medieval Europe, especially during the plague years, the feverfew flower was an essential part of cottage gardens. Local lore said that planting feverfew flowers by the house, especially near the door, would help protect those who were inside from the disease. Interestingly enough, there is some data to support that this may have worked.

AUTHOR'S PERSONAL STORY

As the name already says, feverfew helps people lower their fevers. I've tried this on a lot of patients, and it always works!

24

FIELD GARLIC

LATIN NAME

Allium vineale or *Allium ursinum*

DESCRIPTION

Field garlic, also called wild garlic, grows to approximately two feet tall, and the roots come in the form of bulbs clustered together, each looking like an onion.

The leaves are long and have a hollow tube shape. The leaves grow out in a bunch and resemble the leaves you'd find on chives. They give off a scent that smells like onion. Ensure that whatever you forage has a smell to it, otherwise, it might be a poisonous plant that looks similar.

The flowers are white and pink and are star-shaped. They grow in clusters at the end of stalks that shoot out directly from the root system.

The fruit is called a bulbil. It has an ovular shape and tapers to a tip. The color tends to be green or purple, and it has an oniony smell. These, just like the flowers, grow in clusters. When they're fully mature, they fall to the ground to establish a new wild garlic plant.

HABITAT

They're extremely common in open fields and woods, especially when there's moist soil–such as in a marsh. It loves to grow under trees that only sprout leaves in late spring.

SEASON TO GATHER THE PLANT

You can harvest this herb in the spring and the winter. The flowers bloom in late spring, making this the time to forage for flowers.

PARTS OF THE PLANT TO USE AS MEDICINE

You can use the whole plant for medicinal purposes.

BENEFITS AND PROPERTIES

The wild garlic plant is very healthy for your heart. It lowers blood cholesterol and blood pressure. It also improves blood flow and purifies the blood. People at risk for heart attacks and strokes may find that wild garlic is exactly the solution they've been looking for.

Your digestive system can also benefit from using this plant. It cleanses out the digestive system and improves digestive function. Part of the cause for this is that wild garlic improves bile production and urination, allowing you to break down food better and expel waste better. Cramps in the stomach and intestines can be reduced with wild garlic, and it can treat obesity and obesity-related issues.

Field garlic can also help clear up the skin.

Immune system function improves when you use wild garlic. This has the secondary effect of reducing the likelihood of getting colds, fevers, sore throat, and the flu.

MEDICINE PREPARATION

Tea: Chop up a tablespoonful of wild garlic. Pour over boiling water and let it steep for a few minutes. Strain, add anything you would like to use to adjust the flavor, and drink.

Macerated oil: Simply put a few handfuls of fresh or dry wild garlic (crushed or cut into small pieces) into a glass jar and cover it with your carrier oil of choice. Seal it tightly and place it in a sunny spot. Strain out the plant matter, and add new wild garlic to the oil once a week for three weeks. After this, it'll be ready for use.

In food: This plant is very easy to incorporate into everyday cooking. It ranges from a chive-like flavor for the leaves, to an onion-like flavor for the bulbs. It's generally chopped or mashed when added to fresh dishes and salads, or chopped up and cooked into hot dishes. It's particularly good for adding to homemade pesto.

USING THE HERB

Tea: You can have a cup of tea per day. It's effective for long-term health, especially for the heart and blood. You can use dried wild garlic for this, in which case you would also use about a tablespoonful of the plant.

Macerated oil: I like the oil because you can use the dried version of the plant for this purpose. I also like that you can use it for topical application.

In food: You can freeze wild garlic, allowing you to use it

in a fresh state even when it's not available for foraging.

SIDE EFFECTS AND WARNINGS

Allergies are something you always need to watch out for when using wild plants, including wild garlic. Testing out a small amount before going the whole hog can prevent you from getting a full-on allergic reaction.

Having bad breath goes part and parcel with eating wild garlic, as it would with your average store-bought garlic. You might also feel a bit of a stomach upset on occasion. Heartburn and flatulence also occur when you have too much at a time—and it's easy to overindulge because it is a lovely addition to dishes for flavor.

Avoid wild garlic if you're on blood thinners. The blood-related benefits associated with wild garlic might interfere with your medication.

FUN TIPS AND FACTS

The plant's common name is 'bear garlic'. It comes from the belief that bears ate wild garlic to regain their strength after a long winter's slumber. This can be said also about man; if we eat field garlic it strengthens our bodies.

AUTHOR'S PERSONAL STORY

My family and I have always cooked a lot with it when it's in season. Its taste is delicious and without realizing it, it boosted our bodies' strength and prevented us from catching several annual diseases. Oftentimes, when we're healthy and live a holistic life, we forget about being sick. Now and then, I take a conscious moment to be grateful for the herbs that constantly promote my health and well-being.

25

GARLIC MUSTARD

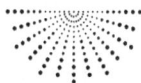

LATIN NAME

A lliaria petiolata

DESCRIPTION

It's a short plant, growing between one and three feet tall, and it has a green to pale gray color.

It has a white taproot, and under the base of the plant, the root has an s-shape.

There's a single, slightly hairy stem that grows from the base of the plant to the flower head. The stem is slender, and the leaves branch out from it.

Its leaves grow in an alternating pattern. They have rough, toothed margins. The leaves have a heart or triangular shape. The size of the leaves ranges between two and three inches long, and two and a half to three inches wide. The leaves have petioles that attach to the stem. There's a distinct garlic scent to the plant that can be quite strong. Note that in the first year, the plant has dark green leaves. These leaves are arranged in the pattern of a rosette and have a kidney shape. These rosette leaves also have large teeth.

The flowers form a clustered flower head called a raceme. The individual flowers in the head produce four white petals that are about a fifth of an inch long. These petals have a blunt or flat tip, and the overall shape of the individual flower is that of a cross or a plus.

The fruit is a silique, which is a slender seed pod that splits open to release its shiny black seeds. The silique grows erect and has four sides. Its coloration turns brown when it's close to releasing its seeds.

HABITAT

Garlic mustard often grows next to hedges and bushes. It generally prefers a colder environment, normally with well-draining soil in the form of loam or sandy soil.

SEASON TO GATHER THE PLANT

If you want to use the flowers, April to June is the optimum time for harvesting these. Otherwise, you can harvest the leaves any time from March to September. It's better to gather the plant before it's at full maturity. The flavor is mild and pleasant, tasting like both garlic and mustard. Note that it lives for two years then dies.

PARTS OF THE PLANT TO USE AS MEDICINE

You can use the leaves, the flowers, and the fruit.

BENEFITS AND PROPERTIES

Garlic mustard is very effective for tending to wounds, bruises, sores, and skin ulcers. It has an antiseptic effect that gives it the added benefit of warding off unwanted microorganisms around or in these. It can also be used topically to relieve muscular cramps, especially in the feet.

It's effective for warding off colds and related issues, such as head congestion and coughing. It can get your body to sweat, which helps get your body temperature back to where it should be when you have a fever or you're overheating. If you have bronchitis, try applying some to your chest.

You will find that it can assist with kidney stones.

As for babies, it can assist with colic. Just consult your baby's medical professional first.

MEDICINE PREPARATION

Fresh: Chop it up and use it in your homemade pesto or your homemade salads.

Poultice: Chop up fresh garlic mustard and mash it up to release any juices. Then add a bit of water so that it's not as thick. Add a bit of honey or soap for extra antibacterial protection and to help it stick to the sore or skin. It needs to be applied quite thick–about half an inch thick.

Oil: put about three handfuls of garlic mustard in a jar and add a carrier oil of your choice so that it covers the plant parts. Seal the jar and leave it in a sunny spot for two weeks, then strain it. You can now use it for topical or food purposes. I go the extra step of melting in a bit of coconut oil and beeswax to make it thick. This way it can be rubbed into the skin as an ointment for bruises and sores.

USING THE HERB

Fresh: It has a tasty flavor that can lend an extra edge to your food. Just don't use too much at once.

Poultice: The poultice is perfect for sores and related issues.

Oil: The oil can be made with fresh or dry garlic mustard, making it perfect for parts of the year when the plant isn't available. The oil will last you a few months.

SIDE EFFECTS AND WARNINGS

Don't take large quantities of garlic mustard because there are trace amounts of cyanide. In small quantities, this won't affect you, but in large quantities, it will.

FUN TIPS AND FACTS

Garlic mustard is invasive and can be harmful to local butterfly species. So don't hesitate to harvest it to your heart's content. There's no need to sustain its population unless you want to cultivate it.

AUTHOR'S PERSONAL STORY

I love using garlic mustard to cook the best blue cheese pasta! It's a treat that I sometimes enjoy, even though I mostly avoid gluten.

GERMAN CHAMOMILE

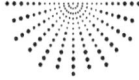

LATIN NAME

Matricaria recutita

DESCRIPTION

There is one basal stem, splitting into multiple light green stems that grow up to the flowers. There is often a light covering of hair along the stem. It's approximately three feet at maximum height but is generally closer to 20 inches high.

Leaves are normally delicate and look similar to fern leaves. They're bipinnate and arranged alternately. Crushing the leaves gives off a smell similar to that of apples.

The flowers have a raised and rounded yellow center. This is surrounded by white petals that span about an inch across. The base that connects the stalk to the flower is hollow.

The fruit of this plant consists of seeds that can be found

on the central yellow part of the flower when it dries. The seeds are very small.

HABITAT

It grows well in temperate areas but can grow in slightly colder areas too. It prefers direct sun and soil that's not too nutritious.

SEASON TO GATHER THE PLANT

The flowers can be harvested from May to October. When you pluck a flower, normally another will start growing to replace the plucked one.

PARTS OF THE PLANT TO USE AS MEDICINE

The flowers are the best part to use for medicinal purposes, especially for topical use. The leaves and stems also are effective. Don't hesitate to use them as well.

BENEFITS AND PROPERTIES

German chamomile is great for oral issues. Use it when you have an abscess, gum inflammation, or a sore throat. Use it as a gargle for best results.

It works on eczema and psoriasis to reduce inflammation and scaly skin. Acne breakouts can be reduced too. When you get light burns on your skin, use it to soothe the skin while it's healing.

Internal benefits include relieving stomach ulcers and inflammatory bowel disease. It can also help you overcome a chest cold. Your stomach issues can be eased with the herb as well. When you have an upset tummy, indigestion, diarrhea, or cramps, this is the herb to use.

For mental health, German chamomile can reduce anxiety. Insomnia can be overcome with this herb as well, so drink a cup of its tea before bed if you have difficulty sleeping.

With children, it can be used to reduce the itching of diaper rash and chickenpox. In oral applications, it can also provide colic relief.

MEDICINE PREPARATION

Tea: Use two tablespoons of dried German chamomile. Pour over a cup of boiling water and let it steep for 10 minutes. Strain and drink.

Bath: Throw a few handfuls of dried German chamomile into your bath and soak in the bath.

Oil: Dry the flowers and crush them up. Then steam them, catching the steam on a pot lid turned upside down. After a few hours, switch off the stove and let it cool. The oily residue on the surface is what you're looking for. Remove it and put it in a dark glass bottle. The oil is removed most easily by placing the pot in the fridge for a few hours once it's cool enough. This allows the oil to congeal, making it easier to skim.

Ointment: Use about five drops of the oil and two tablespoons of coconut oil. Mix them together thoroughly, and apply to affected areas.

USING THE HERB

Tea: You can have up to three cups a day. You can also use it as a gargle

Bath: This helps with inflammation, scaly skin, burns, cuts, and body acne.

Oil: The oil can be added to water (about five drops per glass of water). This can then be drunk to reduce inflammation and viral or bacterial conditions.

Ointment: This is especially effective when applied to acne or acne-prone skin.

SIDE EFFECTS AND WARNINGS

It can cause an allergic reaction in some people.

Vomiting is a potential side effect, but almost always only if you've had a large amount in a short time.

It can cause sleepiness, which is great when you're trying to have a restful night but not so great when you need to drive on a long road or you need the energy to stay awake. Further, avoid it if you're on sleeping pills or sedatives because it could make these stronger than they already are.

German chamomile can also exacerbate asthma.

If you've had cancer linked to high estrogen levels, then avoid this herb, or consult your doctor/oncologist before using it.

If you're pregnant, don't use this herb. You could be putting yourself at risk of miscarriage.

If you're going to have surgery within two weeks, avoid German chamomile. It could cause extra bleeding. In a similar vein, avoid German chamomile if you're on blood pressure or blood-thinning medication.

Other medications that could interact with the plant are birth control, hormone regulators, diabetes medication, antifungals, cholesterol medication, or allergy medication. Ask your healthcare provider if you're on one of these and want to use German chamomile.

FUN TIPS AND FACTS

If you infuse chamomile for longer than 10 minutes in your hot water, it has the opposite effect. Instead of making you sleepy and ready for bed, it wakes you up.

AUTHOR'S PERSONAL STORY

My dad didn't know about German chamomile's potential to increase energy, and when I was a kid, he'd make me chamomile tea to make me tired in the evening. I'd always been a very energetic kid. To boost the effect, he'd soak it longer than the herbalist suggested. The funny story is that he then would get irritated because I just didn't get sleepy.

GOLDENROD

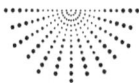

LATIN NAME

S olidago spp.

DESCRIPTION

The stem of this plant is sturdy and grows between three and five feet tall. Most species are shorter (one to three feet) while others are taller (up to eight feet). The stems have tiny spikes on their surface.

The margins of the leaves are serrated, and the surface of the leaf is smooth. They are lanceolate, having a long tapering tip.

The flowers are small—a quarter of an inch in diameter. They grow in a cluster that's long and bound together closely. Their petals are yellow, except in one species called white goldenrod. To me, they look fuzzy.

HABITAT

Some goldenrod species prefer living in mountain and seaside climates. These species normally prefer partial shade to full shade. Most goldenrod, however, prefers direct sunlight and will grow in a variety of areas, such as pastures and fields.

SEASON TO GATHER THE PLANT

Late July to October is the best time to gather the plant. The earlier in this period, the better because the bitterness doesn't come through as much.

PARTS OF THE PLANT TO USE AS MEDICINE

You can use the leaves, the stem, and the flowers.

BENEFITS AND PROPERTIES

Goldenrod can remedy a number of issues including colds, the flu, and allergies.

You can also use goldenrod to treat urinary tract infections and related conditions. Additionally, it increases the flow of urination, allowing your body to expel waste better.

On the skin, this plant can help with eczema and swelling. Other topical benefits include the reduction of pain and muscle spasms. Joint pain is especially susceptible to goldenrod medicine.

A benefit specifically for men is that goldenrod can aid in preventing an enlarged prostate.

MEDICINE PREPARATION

Tea: You'll need about two teaspoons of dried goldenrod. Pour over a cup of water and let it steep for 15 minutes. Strain and drink.

Poultice: Use a handful of finely chopped fresh goldenrod and a cupful of coconut oil. Melt the coconut oil on low heat in a pot and mix in the goldenrod. Pour it into a container that you'll be able to close and let it cool until it's congealed. Apply a decent amount to any affected areas of your skin or muscles.

USING THE HERB

Tea: you can take the tea up to four times a day.

Poultice: keep the poultice refrigerated so that it lasts longer.

SIDE EFFECTS AND WARNINGS

It is possible to have an allergy to goldenrod, so of course, you shouldn't use it if you react poorly to it.

Goldenrod is also harmful if you have heart disease or kidney disease that causes fluid retention. Further, if you're on water pills, the herb may cause you to expel too much water since it has the same function as water pills.

Goldenrod can cause increases in sodium intake. For this reason, it's best to avoid the herb if you have blood pressure issues.

FUN TIPS AND FACTS

In mythology, the goldenrod is thought to indicate the source of a hidden spring (or hidden treasure). It's considered a sign of prosperity. If it starts growing near your home, it means that you could be on to a winning streak.

AUTHOR'S PERSONAL STORY

Clients who were dealing with .undereye bags for years came back after a month of taking goldenrod caps with no sign of bags. The ones who didn't believe in miracles now do!

28

HORSETAIL

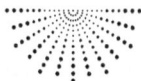

LATIN NAME

Equisetum arvense

DESCRIPTION

The stem of this plant is jointed and hollow. It reaches about one to five feet tall with a few species growing taller. Horsetails also look a lot like rushes. The branches grow out of an underground rootstalk. The stalks are green, and they have gray or black horizontal stripes at their joints.

The leaves form a whorl coming out of the joints of the stem, almost like the spokes of a bicycle wheel. They look like green needles.

The fruit consists of cones that bear spores that have a peach color.

HABITAT

It loves moist areas such as marshes and wetlands, especially when the soil is rich. It can also be found in areas, such as on slopes.

SEASON TO GATHER THE PLANT

Springtime is the best time to harvest this plant. When the leaves turn bright green and turn upward or outward then you won't find a better time to harvest.

PARTS OF THE PLANT TO USE AS MEDICINE

All parts other than the roots are used.

BENEFITS AND PROPERTIES

Horsetail helps with conditions relating to the blood. If you need to reduce the ease with which you bleed, you can use this plant. If you have heavy menstruation, you can also use this herb.

Conditions relating to the joints such as gout and osteoarthritis can be treated with horsetail. The silica content in this herb is a contributing factor to this. The silica content can also contribute to stronger hair and nails, healthier teeth, improved cartilage and ligament health, strengthened lining in arteries and mucus membranes, and stronger bones.

This herb can treat skin injuries.. It can treat frostbite, thereby preventing loss of limbs. Wound care is also greatly enhanced by adding horsetail to your toolbox.

Further, horsetail can aid in weight loss. This is partly due to its ability to reduce blood sugar.

Finally, conditions relating to your urinary tract can be managed effectively when taking horsetail. Incontinence, bladder stones, kidney stones, and urinary tract infections are all remediable by taking this herb.

MEDICINE PREPARATION

Tea: Use a tablespoonful of finely chopped fresh horsetail or dried horsetail. Pour in a cup of boiling water and let it steep in 10 minutes. Strain and enjoy.

Infusion: Chop up a handful of horsetail finely and pour in two cups of hot water. Let it steep for some hours until it's very strong.

TINCTURE: YOU'LL WANT TO USE FRESH HORSETAIL. FINELY chop up a few handfuls and put it in a mason jar. Pour in vodka until the jar is full. Seal it and place it out of the sun for six weeks, shaking it every few days. Strain it and pour it into a dark glass jar for use.

USING THE HERB

Tea: use it consistently for a few weeks to start seeing effects.

Infusion: the infusion can be drunk in smaller quantities–about half a cup per day. You can use it by pouring it directly onto wounds too, or soaking a cloth in it to use it as a compress. It's also great for rinsing out one's mouth or for washing hair. For the hair and nails, it's best to use it as a soak for about half an hour.

Tincture: Use a teaspoon of tincture in a glass of water. Take about one glass a day.

SIDE EFFECTS AND WARNINGS

Using this herb can cause lead to nutrient deficiencies, such as a lack of vitamin B1. Low potassium could be a side effect, specifically if the herb makes you urinate a lot. If you do urinate a lot while taking this herb, you should take a potassium supplement.

Horsetail can also lower your blood sugar. In someone with diabetes, this lowered blood sugar could reach dangerous levels. It's better not to take the herb if you have diabetes or if you're on diabetes medication.

Consult your healthcare provider before taking horsetail if you're on water pills, lithium, or on medication for arrhythmia (a condition in which the heart doesn't beat normally).

Horsetail can make quitting nicotine products harder as it contains nicotine. If you're in the process of trying to get off cigarettes, put the horsetail on hold.

Toxicity is a possible side effect, but it's only likely if taken in large quantities.

FUN TIPS AND FACTS

Horsetail stems have been used as scouring brushes for hundreds of years. They've also been used to buff and shine metal and wooden objects.

AUTHOR'S PERSONAL STORY

I've helped a few recovering addicts detox and boost their kidney function with *equisetum* baths. It's a plant packed with minerals.

29

JAPANESE HONEYSUCKLE

LATIN NAME

L onicera japonica

DESCRIPTION

This plant grows up to 16 feet long. It grows as a creeping or climbing vine with a downy stem. The color of the stem ranges from light brown to reddish-brown.

The leaves are arranged oppositely. They have an ovate shape and a smooth margin (except when they're young, in which case they can have serrated margins). A single leaf is normally between one and three inches long. It has a bright green color.

The flowers are white, yellow, and orange. They tend to start white and turn orange later in the season. They have a tubular shape, looking like there's one petal coming out the bottom of the tube and four attached petals coming out the top of the tube. It gives off a sweet scent, almost like honey and citrus.

Their fruit comes in the form of red or black berries. These are approximately a quarter inch in diameter, and they grow in pairs.

HABITAT

You can find these widespread as weeds, especially in hilly areas that abound with thickets or in the woods. You won't normally find them in very cold areas or very dry areas. In terms of light, it can grow in direct sun or full shade.

SEASON TO GATHER THE PLANT

Harvest the stems in the fall and winter.

The flowers should be harvested in the early morning before they've been opened. They normally flower during June and July.

PARTS OF THE PLANT TO USE AS MEDICINE

You can use the flowers and the leaves. The leaves should be boiled to prevent toxicity.

BENEFITS AND PROPERTIES

It's great at fighting bacteria, inflammation, and toxicity. This helps with conditions such as colds, fevers, and the flu, as well as pneumonia and other lung infections. Bacterial dysentery, enteritis, syphilitic skin disease, and mumps can also be treated with it.

Internal conditions such as hepatitis, difficulty urinating, blood pressure issues, and blood cholesterol levels can be improved with it.

External application can improve rashes (especially if infectious), inflammation, sores, pain, some skin tumors, muscle spasms, and rheumatoid arthritis.

It's also great for improving overall health.

MEDICINE PREPARATION

Tincture: Fill a mason jar with the flowers and cover them with strong brandy or vodka. Seal it off and put it in a dark place like a cupboard for a month and a half. Shake it every few days. When six weeks have elapsed, strain out the flowers and pour the tincture into a dark glass jar.

Extract: put some fresh or dried plant (about two tablespoons) into a jar of boiling water (about two cups). Leave it to soak for a few hours, at which point you can strain out the plant bits.

Bath: bathe with fresh or dried flowers scattered in a tub of hot water. Soak your body in it for half an hour.

USING THE HERB

Tincture: Take about two teaspoons a day max. You can take it directly or mixed in with a glass of water. Mixing it with water makes it less harsh.

Extract: This can be used internally by drinking it. You can use it externally by soaking a rag in it and applying it to your skin as a compress or using it as a wash on rashes, inflammation, and sores.

BATH: Use this method when you have sore muscles or muscle spasms.

SIDE EFFECTS AND WARNINGS

Side effects of Japanese honeysuckle include rash, and diarrhea, both generally linked to honeysuckle allergies.

Avoid this plant when pregnant or breastfeeding unless you've consulted your healthcare provider, and they've given the green light on it.

Don't take it within two weeks of getting surgery because it could make it difficult for your blood to clot. In the same vein, don't take it when you're taking other medication that may thin out your blood, such as aspirin.

FUN TIPS AND FACTS

You can suck on the fresh flower, and it's sweet like honey!

AUTHOR'S PERSONAL STORY

Japanese honeysuckle flowers add a lovely flavor to herbal teas that otherwise taste unpleasant. In the past, I've kept some dried flowers aside in a jar and added them to teas for a bit of extra flavor.

30

LARCH

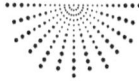

LATIN NAME

L arix

DESCRIPTION

Larch is a large type of coniferous tree, reaching between 50 and 80 feet in height. It has a sturdy stem, supporting branches that spread out wide, with a single tree reaching up to 50 feet wide. The stem can exude resin when injured.

The leaves consist of green needles that come out as a cluster from a bud. These needles go yellow in the fall then fall off during the colder months. The needles are short, reaching about an inch in total.

The fruit is a cone, with each cluster of leaves producing a single cone. These cones also fall off during fall, as the needles do. Cones are green, yellow, red, or purple, ripening to brown after the pollination season. These cones vary in size from species to species.

HABITAT

This tree generally prefers colder temperate areas or cold areas. They're most common in mountainous or hilly regions. A moisture-rich area is their ideal setup.

SEASON TO GATHER THE PLANT

April to May is when the tree buds. This is the best time to gather buds and needles. The inner bark should be harvested in spring when there is a lot of sap flowing through the tree. Just remember to strip vertically and not to remove too much bark, otherwise, the tree will be damaged permanently.

PARTS OF THE PLANT TO USE AS MEDICINE

The bark can be used as well as the resin. Shoots and needles can also be used.

BENEFITS AND PROPERTIES

It has a strong immune-system-boosting effect. This aids in conditions such as the flu, colds, and ear infections.

There is some evidence that it can prevent the growth of liver cancer cells.

Larch increases the number of healthy bacteria in the intestine. This can have a positive effect on one's digestive tract as well as overall health but doing such things as lowering cholesterol.

For brain health, there is evidence that larch can effectively prevent Alzheimer's disease.

Orally, it can be used to treat gum issues and to alleviate a sore throat, and topically, it can be used to heal cuts and bruises.

MEDICINE PREPARATION

Tea: Use a teaspoon of the inner bark (dried and crushed). Pour over a cup of boiling water and let it steep for 10 minutes. Strain out the bark with a fine strainer or a cloth, then drink.

Bath: Use a few handfuls of needles or buds and scatter them in a warm bath. Soak for half an hour.

Sap: You can tap a larch tree and obtain its sap. Do this by creating an inch-deep hole close to the base of the trunk and inserting a straw or tube. Drain the sap into a bottle or jar. Seal the hole afterward for the tree's health. Also, don't tap the same tree more than twice every few years.

USING THE HERB

Tea: This is effective for digestive issues. You can have up to three cups a day.

Bath: This bath will make your body feel stimulated and refreshed.

Sap: The sap doesn't last long (two days at the most), so use it fast. The sap works best for arthritis and inflammation.

SIDE EFFECTS AND WARNINGS

Larch can cause flatulence as well as bloating.

Further, its immune system boosting effect can cause undesirable consequences in patients with autoimmune diseases. The medications taken for this purpose may also interact with larch, causing them to be less effective or have negative reactions.

Individuals that have undergone organ transplants should avoid taking larch. It could cause the rejection of a new organ.

Pregnant and breastfeeding mothers should first consult their healthcare practitioner before taking larch.

FUN TIPS AND FACTS

In European folklore, Larch was thought to have protective powers against evil spirits. It was common to plant it on their property and in villages to protect society.

AUTHOR'S PERSONAL STORY

One of my family members has arthritis that has been giving her problems for years. I used larch sap and made it into a soak for her hands. After soaking her hands for half an hour, she said it was the first time her fingers had felt pain-free in years.

LEMON BALM

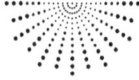

LATIN NAME

M elissa officinalis

DESCRIPTION

The plant comes in the form of a bush that typically grows between 20 and 30 inches high. The stems are thin, green, slightly hairy, and square.

The leaves have an opposite arrangement, and they have an ovular shape. Their margins have rounded teeth, and the veins on the surface of the leaf are rather pronounced. The surface of the leaves feels hairy, and they give off a lemon-like smell and taste similar to lemon.

Flowers grow right against the stem. They're small (half an inch), white, and tubular. There are three rounded petals at the bottom edge of the flower tube, and one at the top. The flower is also slightly hairy.

Lemon balm looks a lot like mint because it's a type of

mint. The main distinguishing factor is the lemony scent and flavor.

HABITAT

Direct sun and fertile, well-drained soil are where this herb shines. It grows in thickets, next to ponds, on floodplains, and in disturbed sites.

SEASON TO GATHER THE PLANT

May to September is when you'll find they provide the best harvest. The flowers specifically are available from June to September.

PARTS OF THE PLANT TO USE AS MEDICINE

The flowers, leaves, and stems are generally used medicinally.

BENEFITS AND PROPERTIES

The main benefit that makes this herb sought after is that it's great for reducing stress and anxiety. This is especially true when it's taken daily. It's also good at making us function better cognitively.

Digestive issues can be managed with this herb too. Indigestion and nausea are common ailments it treats.

Pain can be alleviated well using it. This includes menstrual pain, toothaches, and headaches.

If you have sleep issues, this is a great herb to use. You can get over insomnia by taking it at night before you go to sleep.

Finally, you can use it to heal cold sores.

MEDICINE PREPARATION

Tea: Use half a teaspoon of dried and chopped-up lemon balm. Pour over a cup of boiling water and let it steep for five minutes. Strain, then drink.

Tincture: Chop up fresh lemon balm until you can fill up half a jar with it. Cover it with vodka until there's an inch of vodka above the plant material. Seal the jar and leave it out of direct sunlight for six weeks, shaking the jar occasionally. Strain it and pour the liquid into a dark glass container for storage and use.

Cream: Use a cup of lemon balm leaves (fresh or dried), a quarter cup of beeswax, four tablespoons of glycerine, and two-thirds of a cup of water. Melt the wax on low heat, then put in the water, glycerine, and leaves. Stir it and leave it on low heat for three hours. Strain out the leaves and put the cream into jars. Store it in the fridge.

USING THE HERB

Tea: You can take it up to four times a day.

Tincture: Take up to a teaspoon per day.

Cream: Apply it to affected parts of your skin up to three times a day.

SIDE EFFECTS AND WARNINGS

Pregnant and breastfeeding women may need to avoid it.

The main consideration is that it could interact with other medications and medicinal plants. Particular care should be taken if you're on sedatives, HIV medications, or thyroid medication.

FUN TIPS AND FACTS

Lemon balm is very enticing for bees. Even the ancient Greeks remarked that it was used as a food to draw bees.

AUTHOR'S PERSONAL STORY

I have a friend that had difficulties with sleeping. She'd tried sleeping pills but wanted to get off of pharmaceutical medication because it was causing her to feel odd on a mental level. I suggested lemon balm, and it was just the natural solution she needed.

LINDEN

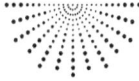

LATIN NAME

Tilia

DESCRIPTION

This is a type of tree with light wood. The bark on the trunk is light gray and there are ridges on it. The bark on the branches is also light gray but smooth.

The leaves are arranged alternately. They come in the shape of an ovular heart with serrated edges and are between four and eight inches long and two to five inches wide. The leaves are green in the spring and summer and yellow in the fall. There is a hairy texture at the bottom of the leaves when they're still small.

The flower is a creamy color, and it gives off a sweet and pleasant smell. It grows as a cyme, which is a cluster of flowers on which the terminal floret flowers first. Linden flowers have a fuzzy appearance.

The fruit of linden is a cluster of seeds that grow down from a peduncle. They are pea-sized and shaped but have a cream color. Each seed has a hard shell.

HABITAT

Lindens prefer a temperate climate, so you won't find them in extremely hot or cold areas.

SEASON TO GATHER THE PLANT

The flowers are gathered in spring. This is also the best time to harvest leaves and bark. Remember to harvest bark responsibly so as not to injure the tree irreparably.

PARTS OF THE PLANT TO USE AS MEDICINE

The flowers are the most valuable parts for medicinal use. The leaves and bark can also be used medicinally.

BENEFITS AND PROPERTIES

Use linden for conditions such as infections, colds, throat irritation, nasal congestion, and coughing for a speedy recovery.

You can also use it on the skin when you have itchy patches.

Linden can help your heart by reducing nervous palpitations, and it can lower your blood pressure.

LASTLY, THIS TREE MAKES MEDICATIONS THAT ARE GREAT sedatives.

MEDICINE PREPARATION

Tea: Dry the flowers and cut or crush them up. Use a quarter teaspoon in a cup of boiling water. Let it steep for fifteen minutes, then strain and drink.

Wash: Make it the same way you would make the tea (but use a teaspoon instead of a quarter teaspoon of the dried flowers). Let it cool until it's lukewarm or cold before use so that you don't burn yourself.

USING THE HERB

Tea: Don't drink more than a cup a day. Use it before bed, ideally, as it is great as a sleep aid.

Wash: Wash any itchy skin with it. For intense itch relief, try soaking a cloth in the wash to make a compress.

SIDE EFFECTS AND WARNINGS

Heart damage can occur if you use the tea regularly, especially if you already have heart disease.

It's also suggested to avoid it if you're pregnant or breast-feeding.

Ironically, it can cause itching, even though it's normally a cure for itching.

Interactions with drugs such as water pills and lithium can occur as well.

FUN TIPS AND FACTS

Germanic mythology (pre-Christianity) considered linden trees to be holy. Dances and ceremonies occurred around the tree as well as community assemblies and judicial meetings. The tree was believed to aid in uncovering the truth.

AUTHOR'S PERSONAL STORY

I once had some itchy skin on my feet after wearing closed leather shoes for a few months. I used a linden wash daily for a week as I was getting concerned about my feet. The anti-fungal quality and the anti-itching quality sorted my feet out beautifully.

MINT

LATIN NAME

M entha

DESCRIPTION

Many mint species have stolons, meaning underground stems that shoot out new stems.

Mint comes in a bush of many stems with leaves attached to them. The stems are normally bright or dark green, and they tend to be pliable. It grows as a bush that's normally 12 to 24 inches high.

The leaves have an opposite arrangement. They are dark green and hairy, with serrated margins (the teeth are round-ed). The veins are normally darker than the leaf and very clear on the leaf's surface.

The flowers are lilac-colored, pink, or white. Depending on the species, the flowers grow on spikes or in whorls. There are normally four petals per floret.

HABITAT

Mint enjoys growing in temperate weather. Full or partial sun and moist soil are just right for this herb. It can grow in all sorts of locations, but next to streams and other bodies of water is most common.

SEASON TO GATHER THE PLANT

You can pick them throughout the year. The best time is just before they flower, but it's not necessary to limit yourself to this timeframe. Try harvesting in the morning before the moisture starts evaporating out of the plant.

PARTS OF THE PLANT TO USE AS MEDICINE

The leaves, stems, and flowers work for medicinal use.

BENEFITS AND PROPERTIES

Mint is commonly used to reduce indigestion, flatulence, and irritable bowel syndrome.

Mint can also improve respiratory function during illnesses like the common cold. This isn't necessarily due to physical changes caused by the mint, but normally more due to subjective perception of those using it. The mint makes you feel like your air passages are clearing up, even if they're not, giving slight relief.

Applying it topically to the nipples can relieve breast-feeding pain. In the same way, it can soothe itchy and irritated skin.

Additionally, mint can improve brain function. It can reduce anxiety, fatigue, and frustration, and it can increase alertness.

It also fights bad breath. There's a reason it's been used in toothpaste.

MEDICINE PREPARATION

Tea: Put a sprig of mint in a cup of boiling water and leave it to steep for 10 minutes. Take out the sprig and drink up.

Oil: Fill a jar with fresh mint and cover the mint with a carrier oil of your choosing. Seal the jar and put it in a spot out of the sun for three days. Strain out the leaves and fill the jar with new mint. Pour the oil back in and seal the jar again. Leave it to soak for another day, at which point you can strain it out again. Pour the oil into a container of your choosing. Store it in a cool and dry place.

Tincture: Fill a jar with fresh mint and cover it with vodka. Seal the jar and put it in a spot out of the sun for six weeks. Shake it daily. When the six weeks have elapsed, strain out the mint and pour the liquid into dark glass bottles for use and storage.

Mouthwash: Mix a teaspoon of tincture with a cup of water. Use it to rinse out your mouth.

USING THE HERB

Tea: Drinking the tea half an hour before a meal helps to keep the digestive system working well.

Oil: Simply rub it onto any affected areas, such as the nipples or dry skin.

Tincture: Mix a teaspoon with a glass of water, then take it orally.

Mouthwash: Use like any mouthwash to leave your breath smelling great.

SIDE EFFECTS AND WARNINGS

Allergic reactions and the triggering of asthma are the most common side effects.

FUN TIPS AND FACTS

Eating fresh mint leaves or putting a drop of its essential oil on your tongue opens your nose and senses like no other plant!

AUTHOR'S PERSONAL STORY

When facing exam stress as a teen, I drank mint tea a few times a day while studying. This helped me stay alert and focused while I was making sure I'd grasped the information I needed to.

MUGWORT

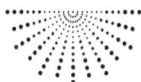

LATIN NAME

Artemisia vulgaris

DESCRIPTION

The stem of this plant is stiff and purple, and it has hairs as well as grooves. The stem goes purplish-red when mugwort is in flower. It can reach up to six feet tall.

It's established in the ground with an eight-inch deep root with rhizomes and rootlets spreading out from it.

Mugwort has leaves that are light (almost white) on the bottom and dark green on the top. The margins are deeply lobed. The leaves can grow up to four inches long, and they're quite thin. The plant also gives off a smell similar to that of sage.

The flowers are small yellow or red wooly flowers. They are arranged as clusters on spikes at the top of the herb.

HABITAT

You can normally find mugwort in temperate areas. The most common areas are along rivers or streams. It's also common on disturbed sites.

SEASON TO GATHER THE PLANT

It's best to harvest the plant in spring before it flowers. But you can still harvest it into the summer when it's in flower.

PARTS OF THE PLANT TO USE AS MEDICINE

The root, stem, blossoms, and leaves can be used medicinally.

BENEFITS AND PROPERTIES

Mugwort is known to assist with stress reduction. It's also known to increase your energy.

Using mugwort can improve digestion as well as blood circulation and liver function. Your urinary tract should also function better, allowing for more regular urinary output.

You can also use the herb to reduce pain, including headaches and muscle aches.

If you have a problem with itching or eczema, this herb is the ideal solution. This is due in part to the antifungal and antibacterial qualities it possesses.

Mugwort can also be used to regulate the menstrual cycle and induce labor.

MEDICINE PREPARATION

Tincture: Fill a jar with dried mugwort and pour in some vodka until it reaches an inch above the mugwort. Close the jar and put it in a dark spot for a month, shaking it now and then for the best result. At the end of the month, strain the mugwort and pour the tincture into a dark glass bottle.

Tea: Use half a teaspoon of dried mugwort. Put it in a cup of water and let it steep for 10 minutes. Strain out the mugwort and drink your tea.

Oil: Fill a jar with dried mugwort and pour in carrier oil until it's level with the mugwort. Close the jar and put it somewhere out of the sun for four days. Shake it once a day, then strain the mugwort and pour the oil into a container.

USING THE HERB

Tincture: You can use up to a teaspoon per day. It's best to have a few drops (about 5) three times per day when using the tincture.

Tea: You should be able to drink it up to three times a day.

Oil: Apply it to your skin as a soothing agent or to sore muscles to reduce pain.

SIDE EFFECTS AND WARNINGS

Allergies to mugwort do occur, and the reaction can be quite strong in some people.

Pregnant women should watch out for this herb since can cause miscarriages. It's also smart to be cautious about using this herb while breastfeeding.

FUN TIPS AND FACTS

Mugwort has been used in alignment with acupuncture for three millennia. It's aged and bound together to form a burning herb called 'moxa.' This complements the acupuncture and is used for its healing power.

AUTHOR'S PERSONAL STORY

I have had several clients who had headaches that were resolved by rubbing mugwort oil into their temples three times a day.

MULBERRY

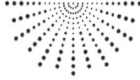

LATIN NAME

orus alba

DESCRIPTION

Mulberries are trees that grow up to 40 feet high, though they're normally smaller. The trunk is short, and many branches stem out of it. There are normally pliable, thin twigs on the branches. The bark is orange-brown, and it has a knotted, fissured texture.

The leaves grow in an alternate arrangement. They grow up to about three inches long and have a toothed margin. They have a heart shape, but some of the leaves lower on the tree can have a lobed appearance or have incisions. The shape can vary, but in general, all the leaves have a downy bottom and a hairy top.

The flowers are catkins that are green, spiky, and small.

The fruit comes in the form of clustered drupes. They tend to be red but can be whitish or purple too. They're normally pink while ripening.

HABITAT

Mulberry trees like direct sunlight best as well as well-draining, nutritious, and mildly moist soil. They like temperate climates but can stand slightly hotter and colder climates.

SEASON TO GATHER THE PLANT

The fruit can be gathered from July to September.

PARTS OF THE PLANT TO USE AS MEDICINE

The fruit and the leaves are good for medicinal uses. Be careful if you decide to use the leaves, though, because a toxic reaction due to the composition of their latex could occur.

BENEFITS AND PROPERTIES

Mulberries have been linked to cancer risk reduction. This is because it helps prevent cells from breaking down due to oxidative stress.

Blood sugar spikes are preventable by taking mulberries. These berries don't prevent the blood sugar increase after a meal altogether, but they reduce the speed with which blood sugar levels rise.

Cholesterol reduction is possible by taking mulberries. The berries assist in breaking down fat and have even shown promise at reducing fatty liver in some experimental studies.

MEDICINE PREPARATION

Fruit: Eating the fresh fruit directly is the simplest form of medication.

Juice: Simply use enough mulberry fruit to fill your container and juice it up. You can also use frozen mulberries to make a smoothie.

Tea: you should use leaves that have been dried and crushed for this purpose. Use a teaspoon of the dried leaves per cup of boiling water. Allow it to steep for five minutes, then strain the leaves and drink the tea.

USING THE HERB

Fruit: Eating the fruit can prevent a blood sugar spike after a meal. The fruit can be frozen to preserve it.

Juice: One glass of mulberry juice is the recommended daily amount.

Tea: you can have this tea three times a day.

SIDE EFFECTS AND WARNINGS

Diarrhea, gas, bloating, or constipation are possible side effects.

Consult your healthcare provider if you want to take mulberries while pregnant or breastfeeding.

FUN TIPS AND FACTS

Mulberry leaves are the sole diet of silkworms. So, without mulberry trees, we wouldn't have silk in the world.

AUTHOR'S PERSONAL STORY

I suggested daily mulberry juice to a man who was having trouble staying awake after large meals due to a sugar spike and crash. The suggestion paid off when he no longer needed to take a nap in his car at work after eating lunch.

MULLEIN

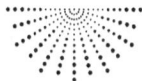

LATIN NAME

Verbascum thapsus

DESCRIPTION

It's a tall plant, growing between five and 10 feet in height. The stem is winged, meaning it has a thin piece of soft green 'skin' running its length. The stem also has a wooly look to it.

There are basal leaves to the plant that grow to a diameter of about two feet. These are pale gray-green. They have an ovular shape and a downy surface texture. The leaves are large near the bottom of the plant and get smaller higher up the plant.

The flowers are yellow and come in the form of tall spikes. Individual florets have five petals with rounded tips that look almost like the round point of a shovel. Smaller flower spikes can grow out of the central flower spike.

The fruit is an oval capsule with a wooly texture, and it

grows on the flower stalk. It's green when the plant flowers, and gradually turns brown.

HABITAT

The plant prefers meadows, woods, pastures, and disturbed areas of land. It likes dry, sandy soil, but it can stand mildly moist soil. Sunny areas are the best for this type of plant.

SEASON TO GATHER THE PLANT

The flowers can be harvested in the summer. The leaves are harvestable year-round and can already be harvested in the first year of growth—choose fresh, young leaves. If you want to harvest the root, it's best to gather in the first year of growth or before any signs of flowering occur during the second year.

PARTS OF THE PLANT TO USE AS MEDICINE

The flowers, leaves, and roots can all be used for this purpose.

BENEFITS AND PROPERTIES

Use the plant when you have an infection, whether it be a middle ear infection, a sinus infection, a urinary tract infection, or a different kind of infection.

Aches and cramps—such as earaches, menstrual cramps, migraines, or toothaches—can all be alleviated using common mullein.

You can use it to remedy a sore throat and coughing. Further, you can use it when you're trying to overcome bronchitis. IT can also aid in the management of asthma and other respiratory problems.

Mullein can also help with urinary control issues, including bed-wetting.

Topically, it is used for skin diseases, bruises, frostbite, arthritis, and rheumatism.

MEDICINE PREPARATION

Tea: Use a teaspoon of dried common mullein (flowers or leaves). Put it in a cup of boiled water and let it steep for 10 minutes. Strain out the plant pieces, then drink them.

Oil: You can make it with a cold or a hot procedure. The cold procedure requires you to use dried leaves or flowers (any amount) and cover it with a carrier oil (I suggest olive oil) until the oil's just above the dried plant bits. Steep it for a week, then strain the oil into a container for use. The hot procedure requires that you place leaves or flowers (dry or fresh) into a pot and cover it with a carrier oil. Heat the oil so that it's on low heat and leave it at that heat for three hours. You can take it off the heat after that and let it cool. After this, you can strain it into a container for use.

Smoking: It's best to dry it and crumble the leaf before smoking it.

USING THE HERB

Tea: You can drink up to four cups a day.

Oil: You can use it internally and externally.

Smoking: Its use is to increase lung health. The smoke it generates has a light quality when inhaled.

SIDE EFFECTS AND WARNINGS

If you're pregnant or breastfeeding, it's best to consult your healthcare provider before using it.

When handling it in the field or while preparing it for use, you may get a rash (due to the hairs on the surface). Not filtering out the hairs in your herbal preparations could also result in mouth irritation.

FUN TIPS AND FACTS

The seeds can make it easier to catch fish. They contain a small amount of a substance that makes fish inactive. Thus, if a fish eats it, you'll be able to get hold of it more easily.

AUTHOR'S PERSONAL STORY

There are a few parents I know who have used common mullein tea while getting their children to overcome bed-wetting. The normal procedure was to get their children to drink some of it an hour before bedtime, to allow the child to go to the bathroom, then put the child to bed.

OAK

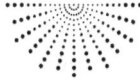

LATIN NAME

*Q*uercus

DESCRIPTION

There are hundreds of oak species. These trees generally grow quite large, growing between 60 and 80 feet (although they can grow as tall as 100 feet or as short as a shrub). Their bark is normally gray when young and becomes white-gray or black when older. Younger trees tend to have smooth bark, whereas older ones tend to have fissures.

The leaves have lobes on them (normally between five and 11). There are two main classifications of an oak tree when it comes to leaves–red and white. The red ones have pointy protrusions on their lobes, whereas the white ones have rounded lobes. A lobe protrudes quite far from the central vein, coming back close to the central vein before the next lobe protrusion starts. The leaves typically grow between five and nine inches long for both classifications. The arrangement of oak leaves is typically alternate. They normally have a green color during the spring and summer, then turn orange, red, and yellow in the fall before they fall off for the winter. Some species, however, are evergreen.

The flowers of an oak tree are normally found on soft, drooping, short stems. They are generally small and green, seemingly insignificant. You may not even realize you're looking at its flowers. However, some species may have flowers that vary widely from this description.

The fruit consists of acorns. Each fruit normally has a peduncle with a cup-shaped green or brown bit. This cup-shaped bit holds a nut that's covered in leathery skin.

HABITAT

Temperate climates are the preferred climate for this type of tree. These trees prefer sun and soil that's well-drained.

SEASON TO GATHER THE PLANT

The bark should be harvested during fall. Acorns should also be harvested during the fall. The leaves should be harvested while they're still green–preferably while they're still young.

PARTS OF THE PLANT TO USE AS MEDICINE

You can use the bark (especially the inner bark), the leaves, the acorns, and gallnuts. (Gallnuts are growths on the tree produced as a result of an infection caused by wasps.)

BENEFITS AND PROPERTIES

Oak is good for the skin because of the contraction it causes in the skin cells. Other skincare uses include reducing inflammation and dermatitis.

It's good for reducing inflammation in the oral region as well. Mouth diseases can be tackled effectively using it. Further, it can be used as a gargle to make a sore throat feel better.

Oak can help your digestive tract function better, and issues related to processing food can be alleviated. It can also regulate urination better, reduce or prevent diarrhea, and assist in breaking down or preventing kidney stones. If used as an enema, it can treat hemorrhoids.

The antibacterial properties of oak can assist in handling vaginal infections.

Lastly, it can help treat rheumatism.

MEDICINE PREPARATION

Tea: You'll need to crush bark into a powder. Add a teaspoon of the bark to a cup of boiling water. Stir it and let it steep for 15 minutes before straining it out. At this point, it should be ready to drink.

Enema: Prepare this in the same way you would the tea, but let it steep for 20 minutes rather than 15.

Snuff: grind inner oak bark until it's extremely fine. You can then snuff it.

USING THE HERB

Tea: The tea is very good at aiding intestinal functions. When cooled, it also makes a good mouthwash or gargle to help with swollen gums, mouth sores, and sore throat. Cooled tea can also be used as a wash to help the skin.

Enema: This is especially good for hemorrhoids.

Snuff: The snuff helps with nose bleeds.

SIDE EFFECTS AND WARNINGS

Intestinal issues could result from using oak bark longer than a few days at a time. It can also cause kidney and liver damage.

You should limit topical use to two weeks. Using it longer could expose you to unwanted side effects, such as eczema and skin irritation..

If you're pregnant or breastfeeding, you should consult your healthcare provider before using it.

Avoid using oak if you have nerve disorders, liver issues, kidney problems, a fever, extensive skin damage, or heart conditions.

FUN TIPS AND FACTS

Oak was representative of the most important god in ancient Greek mythology, ancient Roman mythology, Celtic mythology, Slavic mythology, and Teutonic mythology. The chief god in all these religions was linked to lightning, and ironically, the oak tree is a tree that has a high incidence of getting struck by lightning.

AUTHOR'S PERSONAL STORY

There was a man with hemorrhoids that were preventing him from sitting comfortably. He was embarrassed about it and asked me for advice. I told him how to do an oak enema, and after trying it out for a few days, it went away.

PINE

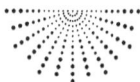

LATIN NAME

P inaceae

DESCRIPTION

This is a tree that can grow quite tall. There are 129 species, so there are many variations to the typical pine tree described here.

The typical pine tree has gray or red-brown bark.

It has needles that grow in clusters of two to five on twigs. They can be anywhere between an inch and 18 inches long.

Pines have seed-bearing cones that come in an egg shape. These cones hang, rather than grow upward (which can be a helpful fact when differentiating from other conifers). The cones release pollen or seeds when they're ripe then fall off.

HABITAT

Where pines grow can vary from deserts to rainforests to cold regions. Most, however, grow in cool, temperate regions. They prefer direct sun, sandy soil that's not too fertile, and soil that's not too wet.

SEASON TO GATHER THE PLANT

The pine is evergreen, so you can harvest it at any point of the year.

PARTS OF THE PLANT TO USE AS MEDICINE

The needles, the bark (especially the inner bark), and the resin of a pine tree can be used.

BENEFITS AND PROPERTIES

Pine can treat fevers, colds, bronchitis, and coughing as well as inflammation in the lungs.

It also helps regulate blood pressure.

It can relieve muscle and nerve pain as well.

When taken with vitamin C, it can improve memory and thinking ability.

MEDICINE PREPARATION

Tea: Take a few pine needles and put them in a cup of boiling water. Let them steep for five to 10 minutes, then remove them. Drink the tea. An alternative to using the needles is using pieces of the bark.

Steaming: Put a bunch of pine needles in a bowl of boiled water fresh from the stove. Put your face over the bowl and throw a towel over your head to prevent the steam from dispersing. Inhale for several minutes.

Bath: Break bark into small pieces and put it in your bathwater.

Wine infusion: Break bark into small pieces and put it in a container/jar with wine. Let the wine become infused with the bark for several hours to a day, then remove the bark. Drink the liquid.

Salve: Use two tablespoons of pitch, four teaspoons of olive oil, and one and a half tablespoons of beeswax. Melt them together on low heat before pouring them into a container. Let it cool and congeal, at which point you'll be able to use it.

USING THE HERB

Tea: The tea can be used as a preventative measure in seasonal changes so that you don't get colds. It can also be used as a wash when it's cooled.

Steaming: Do this to reduce phlegm.

Bath: Soaking in this can alleviate muscle aches.

Wine infusion: The Chinese traditionally use this for joint pain.

Salve: Use this for joint inflammation or insect bites.

SIDE EFFECTS AND WARNINGS

The bark can be toxic if you take larger quantities.

Pine should not be taken orally for long periods..

Allergies can occur even if pine allergy isn't detected in an allergy test.

If you have asthma, it may be a good idea to avoid pine.

Pregnant and breastfeeding women should avoid pine unless their healthcare provider says otherwise.

FUN TIPS AND FACTS

The ancient Romans associated pine with fertility, and they used it in fertility rituals. This is in part because of the phallic shape of a pine cone.

AUTHOR'S PERSONAL STORY

I steam with pine when I have congestion and phlegm. It clears it right up.

39

PLANTAIN

LATIN NAME

P *lantago*

DESCRIPTION

Plantains are low-growers with a basal leaf rosette and spikes with flowers. The leaves are either oval-shaped or lance-shaped and overlap loosely. Veins in the leaves run parallel to each other.

The flowers grow on spikes. They are small and have four transparent petals. The coloration of the spike from a distance is green, yellow, or brown. Note that the flowers don't cover the whole spike, only the upper half of the spike.

HABITAT

Plantain grows ideally where the soil is rocky, sandy, or compacted. It can grow in the shade or direct sunlight. You'll find them in disturbed areas, amongst others.

SEASON TO GATHER THE PLANT

You can gather them any time of the year. Younger leaves are the best.

PARTS OF THE PLANT TO USE AS MEDICINE

The leaves are used for medicinal purposes.

BENEFITS AND PROPERTIES

Plantain has several components that could reduce inflammation in the body. A bonus of this is that you could be protecting your liver health in the process.

It can boost the digestive system and regulate bowel movements. It also works effectively as a natural laxative.

In addition, plantain can help heal wounds, especially due to its anti-microbial characteristics.

It's also good for dealing with insect bites.

MEDICINE PREPARATION

Poultice: To make a good poultice, simply blend or mash some plantain leaves together until you've formed a paste. Apply it to the wound, injury, bite, etc. Once applied, cover it with a bandage.

Tea: Use two teaspoons of dried leaves or two tablespoons of fresh leaves. Put it in a cup of boiling water and stir, then let it steep for 10 minutes. Strain out the leaf bits, then drink up.

USING THE HERB

Poultice: This can only be used once. You'll need to use fresh leaves, not dried leaves.

Tea: You may find that you experience a laxative effect when drinking the tea, so don't have too much. If you do feel a laxative effect, drink water so you don't become dehydrated.

SIDE EFFECTS AND WARNINGS

Allergic reactions may occur. Some allergic reactions are quite strong.

Plantain supplements can have negative effects, such as digestive issues. This is especially the case if the supplements were made using plantain seeds.

FUN TIPS AND FACTS

Plantain can be put on mosquito bites to stop itching or on blisters as a protective barrier.

AUTHOR'S PERSONAL STORY

A family member once had an itchy bug bite. I whipped up a quick poultice and applied it. The poultice provided instant relief.

40

RASPBERRY

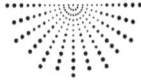

LATIN NAME

Rubus idaeus

DESCRIPTION

Raspberries grow as bushes reaching up to six feet tall. The stem of a raspberry plant is called a cane. It's tall and stiff. Most are prickly, but some species have a smooth cane. In the first year of growth, the cane will be green, and in the second year, it'll be brown. Once it goes brown, it'll be ready to carry fruit then die.

The leaves are shaped like spades and have toothed margins. They have a silvery underside (that's slightly hairy) and a green top. Leaves are normally made up of three to five leaflets.

Flowers vary a lot from species to species, but they normally have five petals with rounded tips. These flowers are normally clustered. Their colors are mainly green or white.

The fruit of this plant is made up of a bunch of drupes. They're soft, with each drupe containing a seed. There are small hairs between the drupes. In terms of shape, a raspberry looks conical but with a blunted tip–and it usually hangs down. There are slight differences in size and color of the fruit from species to species (with the fruit being red, pinkish, purple, yellow, or black).

HABITAT

They are commonly found on river banks, in forest clearings, and in fields. It likes well-drained soil and direct sunlight.

SEASON TO GATHER THE PLANT

The fruit is ready for picking in late June and in July. The leaves can be picked whenever they're green–young leaves being the best.

PARTS OF THE PLANT TO USE AS MEDICINE

The leaves and the fruit are best for medicine.

BENEFITS AND PROPERTIES

Raspberries are good for your blood. They can lower blood pressure and blood sugar (contributing to diabetes management and weight loss). Furthermore, they can improve heart function, thereby combating heart disease. Cholesterol levels can be improved too, and they can reduce your risk of stroke

Eating raspberries is also good for the skin and bones.

MEDICINE PREPARATION

Tincture: Use dried raspberry leaves for this. Fill a mason jar with the leaves and cover it with vodka. Close the jar and put it in a cool space out of direct sunlight. Leave it there for six weeks. Give it a shake every few days to mix up everything in the jar. When the six weeks are up, strain the liquid into a dark glass bottle for use.

Tea: You can use dried or fresh raspberry leaves. Use a tablespoon of the leaf in a cup of boiling water and let it steep for 20 minutes. Remove the leaf pieces (by straining) and drink the tea.

Fresh berries: Simply clean the berries and eat them.

USING THE HERB

Tincture: You can use the same tincture for years without it expiring. Don't use more than a teaspoon per day.

Tea: It tastes similar to black tea, so you can use it as a replacement for your normal tea. As an added bonus, it doesn't contain caffeine.

Fresh berries: The berries are high in vitamins and have very low sugar levels, making them great for snacking.

SIDE EFFECTS AND WARNINGS

People on insulin should be cautious with raspberries due to their ability to lower blood sugar.

If you're on medication to prevent or slow blood clotting, you shouldn't consume raspberries.

Conditions that are sensitive to hormones, such as some cancers, can be exacerbated by raspberry consumption.

FUN TIPS AND FACTS

In the Christian religion, raspberries have often been used to represent kindness in art.

AUTHOR'S PERSONAL STORY

I once had a scare that I might become anemic after visiting a healthcare expert. After taking raspberries multiple times a week for a month, I went back to the expert. The potential crisis had been averted as my blood was carrying iron as it was supposed to.

RED CLOVER

LATIN NAME

Trifolium pratense

DESCRIPTION

Red clover grows between six and 24 inches tall. It has a green, hairy stem.

The leaves on this plant are compound, with three leaflets composing each leaf. The leaflets are between half an inch and two inches long. They're green, and there's a white 'v' shape on the top of the leaf (which is one of the plant's most easily identifiable characteristics). The overall compound leaves are arranged alternately.

The flowers grow on a head that's an inch long and an inch wide. The florets on the head are a half-inch across. They're pink or purple, but they go brown close to winter.

HABITAT

This type of plant likes direct sun and well-drained soil. It grows in pastures, fields, and meadows.

SEASON TO GATHER THE PLANT

It flowers from May to September, and the best time to harvest the flowers is in the latter part of this period, in August and September.

PARTS OF THE PLANT TO USE AS MEDICINE

The dried flower head is the main part of the plant used medicinally.

BENEFITS AND PROPERTIES

Red clover can help prevent heart disease.

Additionally, it can lower the risk of prostate cancer.

Red clover can improve skin and hair health as well. . In fact, there has been a noted connection between red clover and reduced hair loss.

The herb can also reduce symptoms associated with menopause

MEDICINE PREPARATION

Tea: You need two teaspoons of dried flowers. Put it into a cup of hot water (not quite boiling) and let it steep for 15 minutes.

USING THE HERB

Tea: You can have up to three cups a day. You can also let the tea cool and apply it to your skin or hair as a wash.

SIDE EFFECTS AND WARNINGS

If you have hormone-related cancer, then avoid red clover.

Pregnant and breastfeeding women should also avoid it unless their medical professional has advised them that it's okay to use.

Red clover interacts with some drugs, including birth control, medication to reduce blood clotting and blood platelets, rheumatoid arthritis medication, and psoriasis medication. The last two types of medication could interact with red clover in such a way that it causes toxicity in your body.

Avoid taking it within two weeks of surgery, as it could interfere with the medication that is required for the surgery.

FUN TIPS AND FACTS

In folk magic, red clover is used in a ritual bath to attract money and prosperity to the bather and is also used as a floor wash to chase out evil and unwanted ghosts. Four-leaf clovers are famous as a good luck charm believed to protect people from evil spirits, witches, disease, and the evil eye.

AUTHOR'S PERSONAL STORY

A friend of my grandmother's was starting to get concerned because her hair was starting to thin out. I suggested using a red clover wash for the first week of every month (using it to soak the hair for five minutes each of the days in the week). She tried it and her hair loss slowed down dramatically.

SAINT JOHN'S WORT

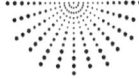

LATIN NAME

H ypericum perforatum

DESCRIPTION

St John's wort is a type of shrub that grows up to 31 inches high and has a stem that discharges a red liquid.

The leaves normally grow in pairs, and they have an ovular shape. They have glands visible on the surface that give off a foxy smell (i.e. a musky and earthy smell).

The flowers of this plant are bright yellow (petals, center, and reproductive organs), are small, and grow in clusters. Each flower has five petals.

The fruit is a soft red capsule that grows in clusters. There are three chambers in the flower that contain small black or brown seeds. It's covered in a sticky material that helps it stick to people and animals that pass by, allowing it to disperse. There are many species of the plant, and some have different types of fruit, such as a single, glossy red berry.

HABITAT

It grows next to hedges, in woods, and in wastelands. It likes rich and well-draining soil that's mildly moist and full sun or partial shade.

SEASON TO GATHER THE PLANT

June to September is the right time to gather Saint John's wort.

PARTS OF THE PLANT TO USE AS MEDICINE

The flowers of the plant are used for medicinal purposes.

BENEFITS AND PROPERTIES

Wounds, burns, and other skin damage is remedied using St John's wort. Muscle pain can also be managed with topical use.

It can also be used for its mental benefits. The plant can help manage ADHD, depression, insomnia, obsessive-compulsive disorder, and somatic symptom disorder. St. John's wort is particularly used to combat depression—specifically mild and moderate depression. It's known to liven your mood due to increases in impulses being sent in the brain.

St John's wort can also alleviate conditions related to menopause.

And internally, it can benefit your lungs and kidneys.

MEDICINE PREPARATION

Tea: You use the flowers (three teaspoons) to make the tea. Boil two cups of water in a saucepan and put in the fresh flowers. Let it steep for four minutes, then remove it from the heat, strain out the flowers, and serve it.

USING THE HERB

Tea: You can use it internally for a range of the conditions listed in the benefits section. It's good to take it before bedtime because it will assist with better sleep during the night. You can also use it externally, after it's cooled down, as a wash or as a compress.

SIDE EFFECTS AND WARNINGS

St. John's wort is known to interact with medications. These include birth control, depression medication, HIV medication, blood cholesterol medication, some cancer medication, heart medications, the blood thinner warfarin, and medication that prevents organ transplant rejection. There are other medications that it interacts with that aren't listed, so consult your healthcare practitioner if you're on regular medication and you want to take the herb.

It can also cause allergic reactions and sensitivity to sunlight.

It should be taken only with the approval of your medical professional if you're pregnant or breastfeeding.

FUN TIPS AND FACTS

Have you ever tried to soak this herb in oil and make an oil extract? You won't believe the color of the result–it's red and looks like blood!

In ancient medicine, herbalists used to pick it on the summer solstice because they believed that that was the day when the herb has the most healing properties.

AUTHOR'S PERSONAL STORY

There have been a few occasions when I wasn't able to fall asleep, so I drank the tea, spent a few minutes reading, then easily drifted off to sleep.

43
SASSAFRAS

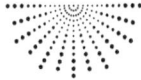

LATIN NAME

S *assafras*

DESCRIPTION

This is a tree that's small in stature, growing between 20 and 40 feet tall. The stem grows between one and two feet across. It has an extensive root system that gives rise to smaller trees. The bark has an aromatic scent and has a red-brown color. It's smooth in younger trees and fissured in older trees.

The leaves have an aromatic scent (almost citrusy), just like the bark. The shape can be triple-lobed (or in some cases, they have five or seven lobes), ovular, or mitten-shaped. They grow between three and seven inches long and up to four inches wide. The leaves have a yellow-green color, and they turn red, orange, purple, or yellow in the fall.

Their flowers grow in two-inch-long clusters at the end of the branches. Individual flowers span one to two inches and are yellow or green. They have six petals and come in the shape of a star.

The fruit of the sassafras tree comes in the form of black or dark blue drupes that grow in clusters. Each drupe contains a seed and grows in a red cup-like peduncle.

HABITAT

Forests, especially at their edges, are good areas to look for sassafras. It likes full to partial sun, and the best soil for this tree is well-drained and loamy.

SEASON TO GATHER THE PLANT

The flowers start growing in early spring, and the leaves are available from spring to late summer. However, the root should be harvested in the winter because all the nutrients will have receded to the roots at that point.

PARTS OF THE PLANT TO USE AS MEDICINE

The roots, the bark, and the leaves can be used.

BENEFITS AND PROPERTIES

It's good for your blood, providing a blood-purifying effect. It also improves blood circulation. You can use it to prevent heart attacks caused by overly viscous blood as it can lower coagulation of the blood.

Sassafras is good for keeping up your liver's health as well.

This tree can have a mild antiseptic effect as well as an immune system boosting effect. This aids in preventing infections, such as those in the stomach.

It can be an effective aphrodisiac. This is partly due to an increase in blood flow to the pelvic area.

Sassafras can also have an energy increasing effect.

This plant is good for treating bruising and swelling. Sassafras is also good for wound care and pain relief (including menstrual pain).

MEDICINE PREPARATION

Decoction: Chop up a piece of root into small pieces. Cut enough to cover the bottom of a small saucepan. Fill the saucepan with water (about three quarters of the way). Bring it to a boil then down to a simmer, letting it simmer until the water darkens. Strain out the root pieces.

Leaves: Wash up a few fresh leaves and rub them on open wounds.

Twigs: Use a fresh sassafras twig and rub it across your teeth and gums as a toothbrush.

USING THE HERB

Decoction: You can drink it as you would drink tea (sweetening and adding milk if you like)–not having more than one cup per day. For topical applications, you can add it to homemade soap to experience its skincare benefits daily. You can also use the decoction as a wash.

Leaves: This has been used by Native Americans to recover faster from wounds.

Twigs: the antiseptic effect this has is great for your oral health.

SIDE EFFECTS AND WARNINGS

Sassafras contains a chemical (safrole) that was labeled as mildly carcinogenic by the FDA in the 1970s. Since then, the same chemical has been linked to anti-cancer properties. Further study is needed to determine its effect.

Overconsumption of sassafras oil can be fatal due to the high quantities of safrole. Sassafras oil needs to be taken in microscopic amounts and should never be given to children.

Miscarriages may result from taking it while pregnant. It's also best not to take it while breastfeeding.

Hot flashes, vomiting, sweating, rashes, hallucinations, and high blood pressure can all be side effects of using this plant medicinally.

When taken in larger quantities, it can make you fall asleep by slowing down your central nervous system. It can also interact with sedative medication, making these medications too strong.

Avoid using it within two weeks of any surgery.

FUN TIPS AND FACTS

Sassafras became popular for its sweet-smelling wood and its healing qualities in the 16th Century. It was exported from the New World to the Old World, being a desirable form of wood at the time. On these journeys to the Old World, sassafras became known as a lucky tree because there were many safe voyages on which the tree was transported.

AUTHOR'S PERSONAL STORY

Once, while I was hiking, I slipped and cut myself on a sharp rock. After rinsing off the cut with water from my bottle, I spotted a clump of sassafras trees. I grabbed a few leaves and crushed them in my hands then rubbed them on my cut. I felt the pain gradually subside. Later, the cut healed much faster than I expected it would.

44

SHEPHERD'S PURSE

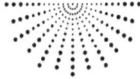

LATIN NAME

Capsella bursa pastoris

DESCRIPTION

The plant grows to about 24 inches max. Most of this height comes from a central stalk that bears the flowers and seed pods.

There are large basal leaves (in a rosette arrangement), then there are smaller leaves on the stalk. The basal leaves die off when the plant flowers. As for their shape, the basal leaves have a long (four-inch), lobed appearance. The lobes on one side of a basal leaf normally correspond to the lobes on the other side–with some being shaped like a round simple leaf, and some having multiple lobes. These leaves are a bit hairy to the touch. The smaller leaves on the stalk are arranged alternately and are slightly toothed.

The flowers of the shepherd's purse are white and appear

in a rounded cluster at the end of the central stalk. The individual flowers are very small—less than half an inch across.

The fruit consists of seed pods that have a heart shape or a triangular shape.

HABITAT

They grow in fields, wastelands, and cultivated areas.

SEASON TO GATHER THE PLANT

Collect the leaves in early spring, before the stalks appear. These leaves will be fresh and pleasant to use. The basal leaves are better to use than the smaller leaves on the stalk. The flowers can be collected from early spring until late fall.

PARTS OF THE PLANT TO USE AS MEDICINE

You can use the shoots, the leaves, the seeds, and the roots. The leaves can be eaten raw or cooked and can be used dry as medicine. The roots can be dried out and used in place of ginger.

BENEFITS AND PROPERTIES

Shepherd's purse can help with heart and blood conditions including (mild)heart failure, mild palpitations, and low blood pressure. Vomiting blood, nosebleeds, and bleeding injuries can also be mitigated.

This plant is also good for the excretory system, and it can treat bladder infections, blood in the urine, and diarrhea.

Menstrual issues can be managed with the herb. If you have cramps while menstruating, a menstrual cycle that lasts very long, or premenstrual syndrome, then you can take it as a remedy.

Headaches and skin burns can also be treated with shepherd's purse.

MEDICINE PREPARATION

Tincture: Fill a jar with fresh shepherd's purse plant matter (cleaned). Pour brandy or vodka in the jar until it's almost at the top. Close the bottle and shake it. Store it in a spot away from direct sunlight for three weeks, and it every few days. Strain the liquid into a dark glass jar once the three weeks are completed.

Food: Use the leaves and the sprouts in your everyday meals. They're good in both salads and cooked dishes.

USING THE HERB

Tincture: Take about 15 drops when needed.

Food: Eating shepherd's purse gives you some of the medicinal benefits of the plant while also feeding your body.

SIDE EFFECTS AND WARNINGS

Overconsumption can cause paralysis, difficulty breathing, and death.

In addition drowsiness, changes in thyroid function, changes in blood pressure, and heart palpitations can result from taking it in normal doses.

Using it while pregnant can result in miscarriage. Consult your healthcare provider if you want to use it while breast-feeding.

Avoid using it within two weeks of surgery because the

lethargy it causes can make anesthesia operate contrary to how it should.

When you have chronic heart problems, you shouldn't use shepherd's purse. It won't assist with your heart health as it would in people without heart conditions. In fact, it could exacerbate any heart conditions you already have.

Shepherd's purse should not be taken by those with thyroid conditions.

This herb can cause those who are prone to kidney stones or have a history of kidney stones to develop more kidney stones.

It will often interact with sedative medication, so don't use it with sedatives.

FUN TIPS AND FACTS

The tincture has many uses, but it's one of the few medicinal plant remedies I know of that can assist with internal bleeding. It's a very good addition to your first aid kit.

AUTHOR'S PERSONAL STORY

I once used the roots (cleaned and finely chopped) in cookies that I was making for my wife while she was going through her menstrual cycle. It served the double purpose of fulfilling the cookie craving and easing the cramping.

45

SOLOMON'S SEAL

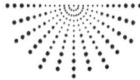

LATIN NAME

P *olygonatum biflorum*

DESCRIPTION

The stem on this plant grows in an arch, reaching knee height. The stem doesn't grow branches, only leaves, and flowers, and it has rhizomes underground.

The leaves come in rows of two and are rather broad. They have an elliptical shape and are arranged alternately along the two rows. They grow between three and eight inches long. The margins are smooth, and the veins are parallel.

The flowers that grow out of the axils of the leaves are either green or white and droop down. They are egg-shaped and have a strong, sweet smell.

The plant produces dark blue berries the shape and size of marbles.

HABITAT

They like moist–but well-drained–soil, so they are commonly found on the banks of bodies of water or in moist forests.

SEASON TO GATHER THE PLANT

The flowers can be gathered in May and June.

PARTS OF THE PLANT TO USE AS MEDICINE

The berries, rhizomes, and leaves can be used.

BENEFITS AND PROPERTIES

The berries act as a laxative in large doses.

Topically, it can be used for a broad range of purposes. It keeps the skin taut, reduces skin redness, reduces swelling, aids in healing bruises, treats hemorrhoids, and helps treat boils.

It also assists with lung disorders and mitigates water retention.

MEDICINE PREPARATION

Tonic: You need fortified wine (enough to cover the other ingredients), Solomon's seal root (eight ounces), and ginger (two ounces). Bring it to a boil in a saucepan, then remove it from the heat as soon as it's started boiling. Let it steep while cooling. Put it in a jar and leave it in a cool, dry place for three weeks. Strain out the Solomon's seal and the ginger. Keep it refrigerated

Tea: Dry, chop and grind up some Solomon's seal root. Put half a teaspoon of this into hot water (just below boiling) and let it steep for 10 minutes. It should be ready to drink.

USING THE HERB

Tonic: This is good for stimulating your digestive tract before eating. Have one ounce before a meal, thrice daily.

Tea: This can also be used externally For this, you'll need to let it cool down, then you can apply it to your skin as a wash or compress. Drink three cups a day at the most.

SIDE EFFECTS AND WARNINGS

People with diabetes should beware of this herb. It could decrease blood sugar, requiring you to monitor your sugar levels. If you're on medication for your diabetes there may be drug interactions.

Pregnant and breastfeeding women shouldn't use it without approval from their healthcare provider.

When going for surgery, don't take it for two weeks prior. The drop in blood sugar that the herb can cause may make medication used in the surgery operate differently from how it's supposed to.

It can upset the stomach, causing diarrhea and vomiting if not used in moderation

FUN TIPS AND FACTS

King Solomon of the Bible is said to have placed his seal on this herb due to the high value he placed on it.

AUTHOR'S PERSONAL STORY

I once accidentally had too many of the berries and got some diarrhea while wildcrafting. Needless to say, the berries are great for cleansing your system.

46

STINGING NETTLE

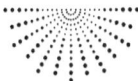

LATIN NAME

Urtica

DESCRIPTION

Stinging nettle grows between two and five feet tall.

The leaves are arranged alternately. They have serrated margins and pointed tips, and their shape is a combination between lanceolate and a heart shape. The leaves of this plant have hairs that sting you, causing itchiness, rash, and pain. The itching comes from histamine and acetylcholine injected through the stinging hairs.

HABITAT

Full or partial sun and nitrogen-rich, nutritious soil is the best for stinging nettle. It's most commonly found near rivers.

SEASON TO GATHER THE PLANT

Don't harvest it after it's started blooming as it can cause kidney damage, whereas before this it can't. Early spring is a good time. Cover yourself while harvesting it because it hurts when it stings.

PARTS OF THE PLANT TO USE AS MEDICINE

The leaves are the main part of the plant used as medicine.

BENEFITS AND PROPERTIES

It's great for managing allergies and hay fever. This is because it contains antihistamine and anti-inflammatory chemicals.

Topical application can reduce pain in joints, tendons, and muscles and pain due to sprains. It can alleviate gout and arthritis as well. It's also good for pain and irritation from insect bites. Additionally, it helps with eczema.

Stinging nettle can help with urinary tract infections, other urinary conditions, and enlarged prostates.

MEDICINE PREPARATION

Tea: Put a handful of leaves in a cup of boiling water, then let it steep for 10 minutes. Afterward, remove the leaves, then drink the tea.

Infusion: The infusion is made the same way as the tea, with the added step of overnighting the steeping. Use this as a compress to affect the external benefits associated with stinging nettle.

USING THE HERB

Tea: Adding mint gives a nice flavor to the tea.

Infusion: Extra infusion that you don't want to drink can be thrown over your potted or garden plants. It's good for them.

SIDE EFFECTS AND WARNINGS

Using it medicinally can cause hives, a rash, diarrhea, sweating, a stomach upset, or fluid retention. However, these aren't commonly experienced.

Stinging nettle can alter the menstrual cycle and cause miscarriages.

It can also interfere with diabetes medication and cause reactions in diabetic patients by raising blood sugar.

Additionally, it is known to interact with medications and with other medicinal plants

Ask for your healthcare provider's guidance before using it if you have kidney or bladder issues.

FUN TIPS AND FACTS

Nettle can be used to make strong cloth. You'll need to learn the processes involved in making it to create your own, but the finish of this cloth is rather durable.

AUTHOR'S PERSONAL STORY

Once, while harvesting stinging nettle, I got pricked by the nettle and had a nasty red spot. Ironically, when I got home, I used a stinging nettle compress to alleviate the pain. So, it can be both the cause of pain and the solution to pain.

47

SUNFLOWER

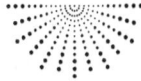

LATIN NAME

Helianthus

DESCRIPTION

Sunflowers are plants that are used extensively in the food and natural medicine industries. There are many species, and there are large variations between some of the species. Characteristics that are common between most of the species include fast growth (up to a few inches a day in spring), a late summer blooming season.

It has a central stem that's upright and stiff. It's usually hairy.

The leaves are large and are hairy to the touch. The shape varies, with some species having heart-shaped leaves, some having oval leaves, and some having leaves that are shaped like spikes. Most have serrated margins.

The flowers have a big disk in the middle, with larger

petals coming off the edges of the disk. The central disk is covered in many tiny florets. The tiny florets each have five petals, five stamens, and a large pistil. The flower changes direction through the course of the day to follow the sun.

The fruit of a sunflower is its seeds. Each of the florets on the central disk is replaced by a single seed. They're shorter than half an inch and covered with a thick husk. The husk is gray, black, and white–streaked vertically. The seed in the middle of the husk is a gray-tan color, and it's shaped like an oval with one pointy end.

HABITAT

Sunflowers grow in prairies and open areas. The climate can vary. Soil should be nutritious and too wet, and sunlight should be direct for long periods of the day.

SEASON TO GATHER THE PLANT

It should be harvested when the flower starts drying out. The base of the flower will be transitioning from green to yellow to brown. This is usually from September to October.

PARTS OF THE PLANT TO USE AS MEDICINE

The seeds are used medicinally.

BENEFITS AND PROPERTIES

Sunflower seeds are good for your heart. They protect against heart attacks, strokes, and high blood pressure by blocking an enzyme that constricts your blood vessels.

It contributes to lowering blood sugar, which can help with managing combat type two diabetes.

Sunflower seeds can also reduce inflammation.

MEDICINE PREPARATION

Oil: Blend the seeds until a paste is formed, adding a pinch of water. Place the paste in a pan at 300 degrees Fahrenheit. Keep it on, and stir it occasionally for 20 minutes. Place it in a fine strainer or cloth over a container for a few hours. The oil will drip out.

Seeds: eating the seeds directly contributes to a range of health benefits.

USING THE HERB

Oil: Heating the oil will reduce the efficacy, and Using shop-bought sunflower oil won't have the same benefits.

Seeds: simply eat an ounce a day to experience the benefits long-term. They can also be added to meals, salads, and baked goods.

SIDE EFFECTS AND WARNINGS

Some people may experience an allergic reaction to sunflower seeds and oil.

The seeds have trace amounts of cadmium, a heavy metal that can cause kidney damage. If you eat very large amounts for a long period, you could experience this side effect. That said, if you eat reasonable quantities, like an ounce per day, this won't be an issue.

There are many calories in sunflower seeds and oil, so overconsumption can lead to weight gain.

In addition, constipation is a common side effect of eating a lot of the seeds in one sitting.

FUN TIPS AND FACTS

Sunflowers don't just look like little suns. They also face the sun during the day, changing the direction of their flowers. Thus, they're appropriately named.

AUTHOR'S PERSONAL STORY

I have used natural sunflower oil on occasion to reduce inflammation in parts of my body. It worked like a charm.

TEASEL

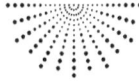

LATIN NAME

Dipsacus fullonum

DESCRIPTION

This plant is quite tall, growing up to six feet high. This is due to the hollow flower stems. These stems are prickly and green.

The leaves grow as a basal rosette at first. The basal leaves have a wrinkly surface, and the margins have rounded protrusions, like the lips of a clamshell. The bottoms of the basal leaves have spines along the midrib. Meanwhile, the leaves on the flower stalk are long, thin, and spiky, and they have a rainwater-holding cup at their base. These leaves have a white line running down the middle and a serrated margin. The leaves.

It grows a seed head that's spiky and shaped like an egg. This is situated at the top of the flowering stem and is about four inches tall. There are light purple or white flowers that grow in rings on this head.

HABITAT

They grow among clumps of trees and in pastures. Teasel can grow in most varieties of soil, so long as it's moist, and it likes the sun.

SEASON TO GATHER THE PLANT

The flowers are available to harvest from July to August. The seeds are ready to harvest from August to October. The root should be harvested in early fall.

PARTS OF THE PLANT TO USE AS MEDICINE

You can use the roots, flowers, and leaves.

BENEFITS AND PROPERTIES

Teasel is good for skincare and preventing skin diseases. It helps with acne, small wounds, warts, red skin, scaly skin, and itchy skin.

It can help with your excretory system, helping achieve more regular urination.

Teasel helps regulate liver health. Among the conditions it alleviates are liver obstructions and jaundice.

This herb can improve digestion. It can stimulate your appetite, strengthen your stomach, and ease stomach aches.

Two other benefits are that it can increase perspiration and it can alleviate arthritis.

MEDICINE PREPARATION

Tea: You can use the root or the leaves. Using the leaves requires about a teaspoon of dried leaf in a cup of boiling water. Steep the leaves for five minutes before straining. The root tea requires a teaspoon of dried and finely cut root per cup of water. Steep it for 10 to 15 minutes before using.

Infused oil: Fill a jar three quarters with dried or fresh teasel, whether it be the leaves, the flowers, or the roots. Cover it with the carrier oil of your choice to the top of the jar. Close the jar, shake it, then leave it in a cool, dark space for three days. Shake it every day. Strain it out after the third day.

Ointment: Use three quarters of a cup of infused oil, a third of a cup of aloe gel, and two thirds of a cup of water. Mix it well, then it'll be ready to use.

USING THE HERB

Tea: You can drink up to four cups of leaf tea or two cups of root tea per day. Other uses include topical use as a wash (especially good for acne) and as an eyewash.

Infused oil: Keep it stored in the refrigerator. You can rub it into your skin for external conditions. For internal use, you can take a spoon directly, or you can drizzle it over your food.

Ointment: The ointment is one of the best folk treatments you can use for warts.

SIDE EFFECTS AND WARNINGS

Pregnant and breastfeeding women should consult their medical practitioner if they plan on using teasel.

It interacts with drugs that are used for nerve conditions, eye conditions, and Alzheimer's.

FUN TIPS AND FACTS

A few drops of teasel tincture can sniff out Lyme disease bacterial cells where they hide burrowed in joints

AUTHOR'S PERSONAL STORY

I once used some leftover teasel tea to wash out the eyes of one of my nephews when they were red and swollen from rubbing. It calmed down the swelling in half an hour, and he was much happier than before.

VALERIAN

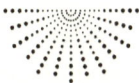

LATIN NAME

Valeriana officinalis

DESCRIPTION

Valerian grows up to six feet tall, but normally only stands three feet tall. The plant's stems are hollow and straight.

The leaves have an opposite arrangement. They're pinnate, with six to 11 pairs of leaflets per leaf. The margins of these leaflets are toothed. The leaves have a slightly hairy bottom surface and an elongated oval shape, tapering to a pointed tip.

Flowers on this tree grow in clusters that are shaped almost like umbrellas. Florets are small (about a tenth of an inch) and are white, light purple, or pink.

The seeds are very small and have a tuft of hair.

HABITAT

It grows well in damp grasslands. It manages well in partial and direct sun and water-rich soil.

SEASON TO GATHER THE PLANT

The root can be harvested at any time of year, but it should be done after rain or when the soil is wet, and it should be done when the plant is at least two years old.

PARTS OF THE PLANT TO USE AS MEDICINE

The root is used for medicine.

BENEFITS AND PROPERTIES

Valerian is used mainly as a sedative and for mental health purposes. It assists with attention deficit hyperactivity disorder, anxiety disorders, and depression. It improves sleep quality, which in turn helps with getting over chronic fatigue syndrome.

Valerian is good for relaxing muscles, easing tremors, restlessness, convulsions, and epilepsy.

It can also alleviate premenstrual syndrome and post-menopausal hot flashes.

MEDICINE PREPARATION

Tincture: Fill a jar halfway with a few handfuls of the root. Cover it with vodka until the jar is full. Seal off the jar, give it a shake, then leave the jar in a cool spot out of direct sunlight for six weeks. After six weeks, strain the liquid into a dark glass bottle.

Tea: Half a teaspoon of dried root is enough. Place it in a cup of boiling water, stir, then leave it to steep for 15 minutes.

USING THE HERB

Tincture: You can have it in water or directly. I suggest a teaspoon per day when you're trying to get the benefits of the herb.

Tea: You can use the tea internally or externally. If used externally, you'll find it works well as a foot bath. You can leave it to steep for a longer period (such as four hours), then pour it into your bathwater. This works as a soak that relaxes muscle spasms.

SIDE EFFECTS AND WARNINGS

It can result in sedation, insomnia, and morning drowsiness.

Mental side effects include uneasiness and excitability in the short term and withdrawal syndrome over the long term.

Valerian can also cause heart issues and liver toxicity.

You may also get mild headaches, dizziness, itchiness, or an upset stomach when taking valerian.

It does interact with alcohol and certain drugs. These drugs include antidepressants, muscle relaxants, and sleeping medication.

FUN TIPS AND FACTS

The ancient Greeks hung bundles of valerian in their homes. They believed that it kept evil entities from entering, so they usually hung the bundles in their windows.

AUTHOR'S PERSONAL STORY

One of my family members used to spend her whole day standing in a shop in high heels. Her feet would be sore all the time. After she started using valerian tea as a foot soak, her feet became a lot less sore.

50

VIOLET

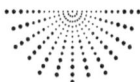

LATIN NAME

V iola

DESCRIPTION

Violet is a low-growing plant. There's commonly a mass of medium-sized leaves growing out of the ground or out of a type of stem that grows horizontally on the ground.

The leaves are shaped like a pointed egg, and they come down at the base to form a bowl. The margins are slightly serrated.

The flowers have five petals, with one of the five functioning as a landing spot for insects that pollinate them. They're either white, purple, or blue. These flowers grow on soft stems that rise out of the ground or from the horizontal stem like the leaves do.

Their fruit is quite small and is either a capsule that explodes with seeds or a berry.

HABITAT

Temperate areas are this plant's preferred climate. It enjoys soil that's rich and well-drained, but moist. Slight shade is best.

SEASON TO GATHER THE PLANT

It flowers in late winter and early spring, so this is a good time to harvest.

PARTS OF THE PLANT TO USE AS MEDICINE

The entire plant can be used medicinally.

BENEFITS AND PROPERTIES

It's used for lung health, bronchitis, coughing, and nasal congestion.

It's also a natural pain medication, which means it can be used to ease headaches and other painful conditions.

This herb can treat your digestive system as well. It reduces inflammation in the intestinal tract, allowing food and drink to flow through better.

It's good for breast health when applied topically. Other topical application benefits include the reduction of swelling and easing skin irritation.

Furthermore, violet has a positive effect on the lymphatic system, letting lymph flow better through your body.

Your quality of sleep will improve while you're on it.

MEDICINE PREPARATION

Syrup: You'll need a cup of sugar, a cup of violet flowers, and a cup of boiled water. Break the petals off, as this is the part of the flower that you'll be using. Pour the cup of water over the petals and let it steep for 24 hours. Then put a pot on the stove, filled halfway with water, and rest a metal bowl on the pot (the bowl should not touch the bottom of the pot). Place the petal infusion in the metal bowl and slowly pour in the sugar, stirring constantly so that it dissolves into the infusion. After the sugar is dissolved into the infusion, strain the mixture into a jar. Let it cool, then it'll be ready for use.

Vinegar: You need a cup of violet flowers and a cup and a half of vinegar (preferably white or rice vinegar). Place the flowers into a jar, then heat the vinegar in a saucepan. Warm the vinegar on low heat, then add it to the jar.. Close the jar and store it in a cool, dry place for three days. After this, strain the flowers out, and the vinegar is ready.

Tea: Use two teaspoons of dried leaves per cup of boiling water. Let it steep for five minutes, then strain it.

Compress: Make it the same way as the tea, but let it steep for 20 minutes before soaking your cloth in the liquid.

Infused oil: You'll need a jar full of violet flowers and leaves, and a carrier oil. Pour the carrier oil into the jar so it covers the leaves and flowers, then close it off. Close the top with a cloth and rubber band. Put it in a cold, dark spot for a month, then strain out the violet leaves and flowers. You can now use it.

USING THE HERB

Syrup: The syrup is very effective for treating coughs.

Vinegar: Have a tablespoon or two of this when your immune system isn't doing well.

Tea: It tastes bitter and like grass, so you may want to adjust the flavor with other plants and herbs or a bit of honey.

Compress: Apply it directly to swollen or irritated skin.

Infused oil: Apply it to the skin or the hair. It benefits both. If used on the skin, it has a soothing quality.

SIDE EFFECTS AND WARNINGS

Pregnant and breastfeeding women should consult their healthcare provider before using it.

Use it in moderate, medicinal amounts. There could be unintended effects if it's used in large quantities.

Allergic reactions are possible.

FUN TIPS AND FACTS

Violets have been used in love potions over the centuries. This is because, in Roman mythology, Venus assaulted a group of girls who then turned into violets after Cupid said the girls were prettier than her.

AUTHOR'S PERSONAL STORY

I've helped someone get over internal inflammation that was preventing them from being able to eat comfortably by getting them to drink violet tea.

51

WILD SARSAPARILLA

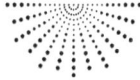

LATIN NAME

Aralia nudicaulis

DESCRIPTION

This plant grows between eight and 20 inches high. It's made up of thin, green stems that come out of the root system and the leaves and flowers attached to the stems

The leaves are alternately arranged compound leaves. The compound leaf splits into three stems that each hold five leaflets. The leaflets are in the shape of a pointed oval they are finely toothed. They're bronze during spring, green in summer, and red or yellow during fall.

Each flower stalk bears three flower clusters. Each cluster is made up of a group of small stalks bearing tiny white flowers. There are twenty to forty flowers in each cluster.

The flowers are replaced by quarter-inch berries that

start green, transition to red, and finally turn dark purple. These have a sweet but spicy flavor.

HABITAT

You can find it next to trees, in bogs, and in woods. It grows in both well-drained and poorly-drained soil, and it prefers partial shade.

SEASON TO GATHER THE PLANT

Gather it at the end of fall when the leaves are starting to fade and the berries have ripened.

PARTS OF THE PLANT TO USE AS MEDICINE

The roots are used for medicinal purposes.

BENEFITS AND PROPERTIES

Internal benefits include protecting the liver from toxins, reducing fluid retention, and preventing kidney disease.

Topical uses include alleviating rheumatoid arthritis, reducing joint pain and swelling, and combatting skin disease.

MEDICINE PREPARATION

Tea: Put a teaspoon of dried and powdered root into a cup of boiling water. Let it steep for 20 minutes before straining and drinking.

Tincture: Fill a fifth of a jar of roots and the rest with vodka. Close the jar and store it in a spot out of the sun for six weeks. Strain the liquid into a dark glass container.

USING THE HERB

Tea: The maximum dosage is a cup per day. You can use it as a soak for sore joints as well.

Tincture: It has a shelf life of several years. You can take half a teaspoon directly, or mixed in with a glass of water.

SIDE EFFECTS AND WARNINGS

A runny nose and asthma-like symptoms can result from breathing in the root's dust.

Wild sarsaparilla can exacerbate kidney disease.

Pregnant and breastfeeding women should consult their medical professionals before use.

This plant can interact with heart medication and water pills.

FUN TIPS AND FACTS

A few Native American cultural groups have used the root as an energy-giving supplement while on journeys.

AUTHOR'S PERSONAL STORY

I've helped a girl lose weight by having her take the tincture so that her body would stop retaining water.

52
WILLOW

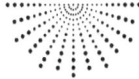

LATIN NAME

*S*_{*alix*}

DESCRIPTION

Willow is a tree that grows many thin and flexible branches that arch out of the trunk and reaches between 30 and 70 feet tall.

Leaves growing on this tree are about a third of an inch wide and one to two inches long. They have finely toothed margins.

Its fruit comes in the form of catkins.

There are many species of willow, and they vary a lot in terms of looks. The above, however, are the descriptive characteristics that are true for most willows. The key identifying factor is the shorter trunk with arched, thin branches.

HABITAT

It prefers full sun and moist ground. It's found close to bodies of water.

SEASON TO GATHER THE PLANT

You can harvest from the willow the whole year long.

PARTS OF THE PLANT TO USE AS MEDICINE

The bark is the main part used for medicinal applications.

BENEFITS AND PROPERTIES

Internal benefits are prevention and recovery from the flu, fevers, and colds. Weight loss is also a possible internal benefit. Pain relief is the main use for willow bark.

Topical benefits revolve around pain reduction. This includes osteoarthritis pain, headaches, menstrual pain, rheumatoid arthritis, spinal pain, muscle pain, gout, and joint pain.

MEDICINE PREPARATION

Tea: Use two teaspoons of crushed bark and stir it into one cup of boiling water. Leave it to steep for 20 minutes, then strain out the bark.

Bark: Simply chew a piece of bark to get the internal benefits described above.

USING THE HERB

Tea: Drink it for internal use.. For external use, soak affected body parts in it or apply it as a compress.

Bark: This may not taste nice, but it's simple and fast pain relief.

SIDE EFFECTS AND WARNINGS

Side effects include allergic reactions, itchiness, rash, headaches, and stomach upsets.

Don't use it if you have a bleeding disorder or kidney disease. It can exacerbate these conditions.

Don't use it on children, as there are possible later brain and liver conditions that could result. More research is required for this.

Don't use it within two weeks of surgery as it could prevent blood clotting.

When pregnant, ask your pregnancy specialist if you can use it or not. While breastfeeding, don't take it because it could lead to damage to your child.

If you're sensitive to aspirin, this isn't the right medicinal plant for you.

It interacts with many medications, so consult your healthcare advisor if you're on medication, and you want to use this.

FUN TIPS AND FACTS

The willow tree can bounce back quickly. It can be cut down yet grow several feet in the following growing season. A piece of twig can be broken off and planted, with new growth occurring easily. These qualities earned it a connection with the concept of immortality in traditional Chinese cultures.

AUTHOR'S PERSONAL STORY

I've chewed on some willow bark after I fell and bruised myself while hiking. The pain faded very quickly after chewing for a few minutes.

53

WINTERGREEN

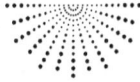

LATIN NAME

Gaultheria procumbens

DESCRIPTION

Wintergreen grows from a network of roots and underground branches that shoot out two to six inches from the ground.

The leaves are arranged alternately. They have an elliptical or oval shape and a length between one and two inches. They're glossy and bright green, and the margins of the leaves have widely-spaced fine teeth with a tiny spine on each tooth.

This plant grows white flowers that have a bell shape. These are waxy and small (about a quarter-inch long). Each flower has five petals.

The fruit is small and berry-like. They're a quarter of an inch in diameter and turn red when matured.

HABITAT

They normally prefer a temperate area. Acidic soil is the most common soil in which this plant flourishes, and it grows best in partial shade.

SEASON TO GATHER THE PLANT

The leaves can be harvested at any time.

PARTS OF THE PLANT TO USE AS MEDICINE

The leaves are used for medicinal purposes.

BENEFITS AND PROPERTIES

Internal uses include pain reduction, treating lung conditions, treating kidney problems, breaking fevers, and improving digestion.

Topical uses include pain reduction, improving skin health, reducing swelling, and killing germs.

MEDICINE PREPARATION

Oil: Put a few handfuls of wintergreen leaves in a saucepan of water. Bring it to a boil, then down to a simmer for three hours. The lid should be upside down on the saucepan the whole time, allowing condensed vapor to drip back into the pot. Let it cool, then place the saucepan in the refrigerator. Leave it there overnight, then scoop up the congealed essential oil (laying as a film on the surface of the water) the following day. Put this essential oil in a dark glass bottle.

USING THE HERB

Essential oil: Put a drop in a glass of water when taking it orally. Put three drops in your bathwater when using it as a soak. Put four drops into eight ounces of carrier oil when using it for topical application.

SIDE EFFECTS AND WARNINGS

The oil shouldn't be taken orally (without diluting it properly). It can cause diarrhea, headaches, stomach aches, ear ringing, nausea, and confusion. These conditions can rise to dangerous levels for children.

Internal side effects can cause intestinal inflammation and stomach inflammation.

The essential oil can irritate if applied directly to the skin.

It interacts with aspirin and Warfarin.

Don't use it if you're pregnant or breastfeeding unless your healthcare provider approves.

People can experience allergic reactions to wintergreen.

FUN TIPS AND FACTS

The scent of the essential oil makes it great for adding to natural cleaning products. You can make your entire house smell fresher by using it.

AUTHOR'S PERSONAL STORY

I've used wintergreen essential oil (breathing in the fumes from the bottle) to get over headaches.

54
YARROW

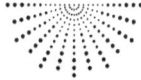

LATIN NAME

chillea millefolium

DESCRIPTION

This is a plant that grows about one to three feet tall.

Its leaves are feathery and grow two to eight inches long. They're arranged in spirals along the stem and are bipinnate or tripinnate. The leaves in the middle and the lower parts of the stem are larger than the ones closer to the end of the stem.

Its flowers are white or pink, and they grow on a flower head. They have a pleasant sweet smell. The flower head has many small flowers (up to 40 sometimes) clustered closely together.

Be careful when identifying yarrow because there are similar plants that are very poisonous.

HABITAT

The plant is commonly found in clearings and meadows, and it likes slightly dry soil.

SEASON TO GATHER THE PLANT

Harvest the leaves in early spring and the flowers from June to August. The roots are best harvested during the fall months.

PARTS OF THE PLANT TO USE AS MEDICINE

The roots, leaves, and flowers can be used.

BENEFITS AND PROPERTIES

Topical uses include toothache relief, hair care, skin cleansing, stanching bleeding, reducing inflammation, and wound care.

There are digestive benefits in the form of stomach and intestinal discomfort relief and appetite stimulation by improving saliva production and stomach acid regulation.

Excretory tract functioning can improve when taking yarrow. It can handle diarrhea, dysentery, bloating, and flatulence.

Women, specifically, can benefit from using it by seeing a reduction in menstrual cramps and pelvic cramps.

Other internal uses include breaking fevers, treating colds, relieving hay fever, and promoting perspiration.

MEDICINE PREPARATION

Tea: Use a teaspoon of leaves or flowers. Put it in a cup of boiling water and let it steep for 20 minutes.

Tincture: Fill a third of a jar with flowers or leaves (fresh) and two thirds of the jar with vodka. Close the jar and put it in a cool, dry place for six weeks. Shake it every few days. Strain out the leaves/flowers.

Bath: Boil a gallon of water in a saucepan then put in a whole yarrow plant. Bring it down to a simmer, then let it simmer for half an hour. Pour this into your bath water, then have a soak.

USING THE HERB

Tea: Have up to two cups per day.

Tincture: Take about half a teaspoon per day.

Bath: You should remain in the bath for a long time so that you can soak up the yarrow.

SIDE EFFECTS AND WARNINGS

It can cause sleepiness, increased urination, and cause skin irritation.

Don't take it when pregnant or breastfeeding. If taken while pregnant, it can cause miscarriage.

If you have a bleeding disorder, you should avoid it.

When going to surgery, avoid using it during the two weeks prior.

Some experience allergic reactions.

It interacts with a range of medications, so consult your healthcare provider if you're on medication and would like to use yarrow.

FUN TIPS AND FACTS

In Greek mythology, the herb yarrow was considered so powerful it was believed to bestow immortality on those who bathed in its waters. According to legend, Achilles was one such hero, having been dipped in the Yarrow-laced water of the river Styx by his mother when he was a baby.

AUTHOR'S PERSONAL STORY

I gave a friend a yarrow leaf to chew when he had a toothache. This helped alleviate the discomfort until he managed to see the dentist later that day.

AFTERWORD

This book describes the most powerful medicinal herbs from the Northeast. It's an in-depth guide into the world of wildcrafting and herbalism including fun stories, facts, and intricate myths about the plants and can be used by new to experienced herbalists.

Take your health into your own hands! Heal truly by getting in touch with nature, our true essence.

Why would you still waste hundreds of dollars yearly on expensive medicine that weakens your body? Go into nature and be your own pharmacist! It's free and in nature, that's where true healing occurs.

If you enjoyed the content of this book, please, take 2 minutes to leave a review. We also have an active Facebook group and we would love to have you so we can learn together.

https://www.facebook.com/groups/northeasthomestead

Let's help humanity rise together!

BIBLIOGRAPHY

5 tips for when you need help identifying a plant. (n.d.). Nature Mentor. Retrieved February 5, 2022, from https://nature-mentor.com/need-help-identifying-a-plant/

6 fantastic reasons for growing and using a comfrey plant. (2019, April 22). Preparedness Mama. https://preparednessmama.com/growing-and-using-comfrey/

10 fascinating and fun cranberries facts from That's It. (2015, November 3). That's It. https://www.thatsitfruit.com/blogs/default-blog/10-interesting-facts-about-cranberries

A guide to garden flower identification: What's in your garden? (n.d.). Plant Snap. Retrieved February 5, 2022, from https://www.plantsnap.com/blog/garden-flower-identification/

Acer negundo. (n.d.). Wild Flower. Retrieved February 22, 2022, from https://www.wildflower.org/plants/result.php?id_plant=acne2#:~:text=Leaf%3A%20Opposite%2C%20pinnately%20compound%2C

Acorn facts for kids. (n.d.). Kiddle. Retrieved March 2, 2022, from https://kids.kiddle.co/Acorn#:~:text=The%20acorn%20is%20the%20fruit

Adamant, A. (2018, May 12). *How to make burdock tincture.* Practical Self Reliance. https://practicalselfreliance.com/burdock-tincture/#:~:text=Chop%20the%20root%20into%20chunks

Adamant, A. (2021, April 5). *15 ways to use borage.* Practical Self Reliance. https://practicalselfreliance.com/borage-uses/

Adriana. (2021, June 24). *Blackberry leaf tea: A herbal remedy for your health.* Backyard Garden Lover. https://www.backyardgardenlover.com/blackberry-leaf-tea/

Agarwal, S. (2021, October 19). *8 amazing mulberry benefits: Make the most of it while the season lasts.* NDTV Food. https://food.ndtv.com/food-drinks/8-amazing-mulberry-benefits-make-the-most-of-it-while-the-season-lasts-1685146

Agrimonia gryposepala (common agrimony, tall hairy agrimony). (n.d.). Native Plant Trust: Go Botany. Retrieved February 7, 2022, from https://gobotany.nativeplanttrust.org/species/agrimonia/gryposepala/

Agrimony health benefits and side effects. (n.d.). Medical Health Guide. Retrieved February 8, 2022, from http://www.medicalhealthguide.com/herb/agrimony.htm

Agrimony: Overview, uses, side effects, precautions, interactions, dosing and reviews.

(n.d.). Web MD. Retrieved February 8, 2022, from https://www.webmd.-com/vitamins/ai/ingredientmono-604/agrimony

Alcantara, S. T. (n.d.). *How to make burdock tea.* Live Strong. Retrieved March 2, 2022, from https://www.livestrong.com/article/52018-make-burdock-tea/

Alder. (n.d.). In Merriam Webster. Retrieved February 8, 2022, from https://www.merriam-webster.com/dictionary/alder

Alder buckthorn. (2021, June 11). RxList. https://www.rxlist.com/alder_buck-thorn/supplements.htm

Alder tincture recipe. (2020, April 3). Cortes Currents. https://cortescurrents.-ca/yulia-kochubievskys-alder-tincture-recipe/

Alder trees plants advantages and disadvantages, side effects and reviews. (n.d.). Review Guts. Retrieved February 8, 2022, from https://reviewguts.-com/alder-trees-plants_2nd/

Alder trees: Leaves, bark, flowers, cones - Identification (pictures). (2021, April 19). Leafy Place. https://leafyplace.com/alder-trees/

American sarsaparilla (spikenard) – Aralia racemosa. (n.d.). Root Buyer. Retrieved March 4, 2022, from https://rootbuyer.com/wild-sarsaparilla-spikenard-aralia-racemosa/#:~:text=The%20rootstock%20is%20har-vested%20in

American Survival Guide. (2018, March 22). *Foraging gear: The tools you need to collect, process and carry natural foods.* American Outdoor Guide. https://www.americanoutdoor.guide/how-to/foraging-gear-the-tools-you-need-to-collect-process-and-carry-natural-foods/

Ancient equisetum. (2015, May 21). In Defense of Plants. https://www.inde-fenseofplants.com/blog/2015/5/21/ancient-equisetum

Andrea. (2019, January 9). *Wild cherry bark & horehound tincture tea recipe.* Frugally Sustainable. https://frugallysustainable.com/cherry-bark-hore-hound-tincture/

Applebaum, G. (n.d.). *Moxa: The burning herb that heals. Fabriq.* Retrieved March 1, 2022, from http://www.fabriqspa.com/the-burning-herb-that-heals/

Baessler, L. (2021, April 6). *What is valerian: How to grow valerian plants in the garden.* Gardening Know-How. https://www.gardeningknowhow.-com/edible/herbs/valerian/growing-valerian-herb-plants.htm

Baldridge, K. (2020, May 8). *How to make an elderberry tincture. Traditional Cooking School.* https://traditionalcookingschool.com/health-and-nutri-tion/make-your-own-elderberry-tincture/

Barth, B. (2018, July 18). *7 smokable plants you can grow that aren't marijuana.* Modern Farmer. https://modernfarmer.com/2018/07/7-smokable-plants-you-can-grow-that-arent-marijuana/#:~:text=Mullein%20(Verbas-cum%20thapsus)&text=Herbal%20Properties%3A%20Mullein%20has%20a

Beaulieu, D. (2021, August 11). *St. John's wort plant profile*. The Spruce. https://www.thespruce.com/st-johns-wort-plant-profile-4772327

Beech. (n.d.). In Merriam Webster. Retrieved February 8, 2022, from https://www.merriam-webster.com/dictionary/beech

Beech. (n.d.). Vild Mad. Retrieved February 11, 2022, from https://vild-mad.dk/en/ingredients/beech

Beech - A guide to its food, medicine and other uses. (n.d.). Eat Weeds. Retrieved February 11, 2022, from https://www.eatweeds.co.uk/beech-fagus-sylvatica

Beech trees: Types, leaves, bark — Identification guide (pictures). (n.d.). Leafy Place. Retrieved February 8, 2022, from https://leafyplace.com/beech-trees/

Benefits of black cherries. (n.d.). Bremner Foods. Retrieved February 11, 2022, from https://www.bremnerfoods.com/health-benefits/benefits-of-black-cherries.html#:~:text=Health%20Benefits%20of%20Black%20Cherries&text=Black%20cherries%20are%20an%20excellent

Benefits of growing medicinal herbs. (2016, March 28). Joybilee Farm. https://joybileefarm.com/7-reasons-make-healing-homegrown/

Berries as symbols and in folklore. (n.d.). New York Berry News, 6(1). Cornell University's College of Agriculture and Life Sciences. Retrieved March 3, 2022, from chrome-extension://efaidnbmnnnibpcajpcglclefindmkaj/viewer.html?pdfurl=https%3A%2F%2Fcpb-us-e1.wpmucdn.com%2Fblogs.cornell.edu%2Fdist%2F0%2F7265%2Ffiles%2F2016%2F12%2Fberryfolklore-2ljzt0q.pdf&clen=100303&chunk=true

Birch. (n.d.). In Merriam Webster. Retrieved February 11, 2022, from https://www.merriam-webster.com/dictionary/birch

Birch fruits and seeds. (n.d.). Tree Guide. Retrieved February 11, 2022, from http://www.tree-guide.com/birch-fruits-and-seeds

Birch Leaf. (n.d.). Mountain Rose Herbs. Retrieved February 11, 2022, from https://mountainroseherbs.com/birch-leaf

Birch perfume ingredient, birch fragrance and essential oils Betula, family betulaceae. (n.d.). Gragrantica. Retrieved February 11, 2022, from https://www.fragrantica.com/notes/Birch-31.html

Birch tree allergen facts, symptoms, and treatment. (n.d.). Thermo Fisher. Retrieved February 11, 2022, from https://www.thermofisher.com/allergy/us/en/allergen-fact-sheets.html?allergen=birch-tree

Birch trees: Types, leaves, bark - Identification (with pictures). (n.d.). Leafy Place. Retrieved February 11, 2022, from https://leafyplace.com/birch-trees/

Birch: Uses, side effects, interactions, dosage, and warning. (n.d.). Web MD. Retrieved February 11, 2022, from https://www.webmd.com/vitamins/ai/ingredientmono-352/birch

Bjarnadottir, A. (2019, February 22). *Mulberries 101: Nutrition facts and health benefits*. Healthline. https://www.healthline.com/nutrition/foods/mulberries#bottom-line

Black cherry. (n.d.). In Merriam Webster. Retrieved February 11, 2022, from https://www.merriam-webster.com/dictionary/black%20cherry

Black cherry. (n.d.-a). Tree Guide. Retrieved February 11, 2022, from http://www.tree-guide.com/black-cherry

Black cherry. (n.d.-b). Natural Resource Stewardship. Retrieved February 11, 2022, from https://naturalresources.extension.iastate.edu/forestry/iowa_trees/trees/black_cherry.html

Blackberry. (n.d.). In Merriam Webster. Retrieved February 11, 2022, from https://www.merriam-webster.com/dictionary/blackberry

Blackberry. (2015, August 6). Kaiser Permanente. https://wa.kaiserpermanente.org/kbase/topic.jhtml?docId=hn-2045002#hn-2045002-side-effects

Blackberry | fruit. (2022). In Encyclopædia Britannica. https://www.britannica.com/plant/blackberry-fruit

Blankespoor, J. (2017, November 7). *Foraging for wild edibles and herbs: Sustainable and safe gathering practices.* Chestnut School of Herbal Medicine. https://chestnutherbs.com/foraging-for-wild-edibles-and-herbs-sustainable-and-safe-gathering-practices/

Blankespoor, J. (2021a, February 4). *The Medicine of Pine.* Chestnut School of Herbal Medicine. https://chestnutherbs.com/the-medicine-of-pine/#:~:text=Pine%20offers%20relief%20in%20sinus

Blankespoor, J. (2021b, April 13). *Violet's edible and medicinal uses.* Chestnut School of Herbal Medicine. https://chestnutherbs.com/violets-edible-and-medicinal-uses/#:~:text=Violet%20is%20cooling%20and%20moistening

Blue vervain. (n.d.). In Merriam Webster. Retrieved February 12, 2022, from https://www.merriam-webster.com/dictionary/blue%20vervain

Blue vervain: Pictures, flowers, leaves & identification | Verbena hastata. (n.d.). Edible Wild Food. Retrieved February 11, 2022, from https://www.ediblewildfood.com/blue-vervain.aspx

Borage. (n.d.). In Merriam Webster. Retrieved February 12, 2022, from https://www.merriam-webster.com/dictionary/borage

Borage leaves. (n.d.). Specialty Produce. Retrieved February 11, 2022, from https://specialtyproduce.com/produce/Borage_Leaves_11921.php

Borage: Uses, side effects and warnings, interactions, dosage, and warning. (n.d.). Web MD. Retrieved February 12, 2022, from https://www.webmd.com/vitamins/ai/ingredientmono-596/borage

Borago officinalis, borage: Identification, distribution, habitat. (n.d.). First Nature. Retrieved February 11, 2022, from https://www.first-nature.com/flowers/borago-officinalis.php#:~:text=Up%20to%2060cm%20tall%2C%20this

Bratianu, P. (n.d.). *The natural healing power of oak trees and acorns.* Off the Grid News. Retrieved March 2, 2022, from https://www.offthegridnews.-

com/alternative-health/the-natural-healing-power-of-oak-trees-and-acorns/

Brennan, D. (Ed.). (2020a, September 17). *Health Benefits of Mint Leaves*. Web MD. https://www.webmd.com/diet/health-benefits-mint-leaves#:~:text=When%20it%20comes%20to%20medicinal

Brennan, D. (2020b, September 18). *Health benefits of raspberries*. Web MD. https://www.webmd.com/diet/health-benefits-raspber-ries#:~:text=They%20provide%20potassium%2C%20essential%20to

Brennan, D. (Ed.). (2020c, November 10). *Health benefits of mulberries*. Web MD. https://www.webmd.com/diet/health-benefits-mulberries#2-5

Buckner, H. (2020, March 20). *How to plant and grow plantain, a culinary and medicinal herb*. Gardener's Path. https://gardenerspath.com/plants/herbs/grow-plan-tain/#:~:text=All%20of%20these%20species%20grow

Burdock. (n.d.). In Merriam Webster. Retrieved February 12, 2022, from https://www.merriam-webster.com/dictionary/burdock

Burdock: Pictures, flowers, leaves & identification | Arctium lappa. (n.d.). Edible Wild Food. Retrieved February 12, 2022, from https://www.ediblewild-food.com/burdock.aspx

Can you juice soft summer fruits like cherries, berries, and peaches? (2017, July 19). Eujuicers. https://www.eujuicers.com/magazine/can-you-juice-soft-summer-fruits-like-cherries-berries-and-peaches

Carroll, J. (2021a, May 4). *Growing a larch tree: Larch tree types for garden settings*. Gardening Know-How. https://www.gardeningknowhow.com/ornamental/trees/larch/growing-a-larch-tree.htm#:~:text=Larch%20trees%20are%20large%20deciduous,flow-ers%20that%20eventually%20become%20cones.

Carroll, J. (2021b, May 10). *St. John's wort plant care: How to grow St. John's wort*. Gardening Know-How. https://www.gardeningknowhow.com/edi-ble/herbs/st-johns-wort/st-johns-wort-plant-care.htm

Carryopsis, J. (n.d.). *Biology of dandelions*. Nature North. Retrieved February 16, 2022, from http://www.naturenorth.com/summer/dandelion/Dande-lion2.html

Catnip. (n.d.). In Merriam Webster. Retrieved February 12, 2022, from https://www.merriam-webster.com/dictionary/catnip

Catnip oil extraction methods, process, techniques. (n.d.). Agri Farming. https://www.agrifarming.in/catnip-oil-extraction-methods-process-techniques

Catnip tea: Are there health benefits? (n.d.). Web MD. Retrieved March 2, 2022, from https://www.webmd.com/diet/catnip-tea-health-benefits#1

Catnip uses, benefits & side effects. (n.d.). Drugs. Retrieved February 12, 2022, from https://www.drugs.com/npc/catnip.html#:~:text=Medicinal-ly%2C%20the%20plant%20has%20been

Chamomile: Matricaria recutita. (n.d.). Edible Wild Food. Retrieved February 24, 2022, from https://www.ediblewildfood.com/chamomile.aspx

Chappell, S. (2019, February 21). *A beginner's guide to making herbal salves and lotions*. Healthline. https://www.healthline.com/health/diy-herbal-salves#TOC_TITLE_HDR_1

Cherney, K. (2022, February 3). *Everything you need to know about borage oil*. Healthline. https://www.healthline.com/health/borage-oil#side-effects

Chickweed. (n.d.). In Merriam Webster. Retrieved February 12, 2022, from https://www.merriam-webster.com/dictionary/chickweed

Chicory. (n.d.). In Merriam Webster. Retrieved February 12, 2022, from https://www.merriam-webster.com/dictionary/chicory

Chicory - edible wild plant - how to find, identify, prepare, and other uses for survival. (2012, May 11). Wilderness Arena. https://www.wildernessarena.com/food-water-shelter/food-food-water-shelter/food-procurement/edible-wild-plants/chicory

Chicory: Health benefits, uses, side effects, dosage & interactions. (2021, November 6). Rx List. https://www.rxlist.com/chicory/supplements.htm

Choosing a location for raspberry plants. (n.d.). Stark Bro's. Retrieved March 3, 2022, from https://www.starkbros.com/growing-guide/how-to-grow/berry-plants/raspberry-plants/location

Christiansen, S. (2022a, January 10). *What is mugwort? Herb related to ragweed used in naturopathic and traditional Chinese medicine*. Verywell Health. https://www.verywellhealth.com/mugwort-benefits-side-effects-dosage-and-interactions-4767226

Christiansen, S. (2022b, February 5). *What Is Horsetail?* Very Well Health. https://www.verywellhealth.com/horsetail-4692253#:~:text=Traditionally%20horsetail%20has%20been%20used

Clark, P. (2013, May 21). *Blackberry sexuality. It's complicated*. The Washington Post. https://www.washingtonpost.com/wp-srv/special/metro/urban-jungle/pages/130521.html

Cleansing with cleavers. (2019, April 8). Botanica Health. https://www.botanicahealth.co.uk/cleansing-with-cleavers/

Cleavers. (n.d.). In Collins. Retrieved February 12, 2022, from https://www.collinsdictionary.com/dictionary/english/cleavers

Cleavers. (2015a, May 23). Peace Health. https://www.peacehealth.org/medical-topics/id/hn-2070002#:~:text=by%20Scientific%20Studies)-

Cleavers. (2015b, May 23). Kaiser Permanente. https://wa.kaiserpermanente.org/kbase/topic.jhtml?docId=hn-2070002#:~:text=by%20Scientific%20Studies)-

Climan, A. (2020, December 21). *What is cleavers (galium aparine)?* Very Well Health. https://www.verywellhealth.com/cleavers-health-benefits-5084341#toc-possible-side-effects

Colleen. (2018, February 2). *Wild violet flower infused vinegar*. Grow Forage

Cook Ferment. https://www.growforagecookferment.com/wild-violet-flower-infused-vinegar/#:~:text=There%20are%20also%20some%20medicinal

Colleen. (2020a, March 26). *How to Make and Use Dandelion Salve*. Grow Forage Cook Ferment. https://www.growforagecookferment.com/how-to-make-dandelion-salve/

Colleen. (2020b, June 15). *Dandelion foraging: Identification, look-alikes, and uses*. Grow Forage Cook Ferment. https://www.growforagecookferment.com/foraging-for-dandelions/#:~:text=Identifying%20Dandelion

Colleen. (2020c, July 28). *Foraging plantain: Identification and uses*. Grow Forage Cook Ferment. https://www.growforagecookferment.com/plantain-natures-band-aid/#:~:text=Identifying%20Plantain

Comfrey information. (n.d.). Mount Sinai Health System. Retrieved February 12, 2022, from https://www.mountsinai.org/health-library/herb/comfrey#:~:text=Comfrey%20roots%20and%20leaves%20contain

Comfrey: Uses, side effects, and more. (n.d.). Web MD. Retrieved February 12, 2022, from https://www.webmd.com/vitamins/ai/ingredientmono-295/comfrey

Common Agrimony, Agrimonia eupatoria - Flowers. (n.d.). Nature Gate. Retrieved February 7, 2022, from https://luontoportti.com/en/t/1025/common-agrimony

Common agrimony: Pictures, flowers, leaves & identification. (n.d.). Edible Wild Food. Retrieved February 2, 2022, from https://www.ediblewildfood.com/common-agrimony.aspx

Common chickweed. (2015, June 5). Michigan State University Integrated Pest Management; Michigan State University. https://www.canr.msu.edu/resources/common_chickweed

Common mallow (malva neglecta). (n.d.). Illinois Wildflowers. Retrieved February 12, 2022, from http://www.illinoiswildflowers.info/weeds/plants/cm_mallow.htm

Common mullein. (n.d.). Woodland Ways Bushcraft Blog. Retrieved March 2, 2022, from https://www.woodland-ways.co.uk/blog/hedgerow-medicines/common-mullein/#:~:text=Key%20Identification%20Features%3A%20Biennial%20up

Common name comfrey (Common comfrey, healing-herb, knit-back, knit-bone, backwort, bruise-wort, slippery-root, asses' ears). (n.d.). Friends of the Wild Flower Garden. https://www.friendsofthewildflowergarden.org/pages/plants/comfrey.html

Common teasel identification and control. (2022, January 7). King County. https://kingcounty.gov/services/environment/animals-and-plants/noxious-weeds/weed-identification/common-teasel.aspx#:~:text=Cut-leaf%20teasel%20has%20irregularly%2Dcut

Common violet. (n.d.). St Olaf College. Retrieved March 4, 2022, from

https://wp.stolaf.edu/naturallands/forest/ephemerals/commonvio-
let/#:~:text=Violets%20are%20flowers%20with%20five

Cramp Bark. (2021, June 11). Rx List.
https://www.rxlist.com/cramp_bark/supplements.htm

Cramp bark. (n.d.). Monterey Bay Herb Co. Retrieved February 15, 2022, from
https://www.herbco.com/c-253-cramp-bark.aspx

Cramp bark (viburnum opulus l.). (n.d.). Health Embassy. Retrieved February 15,
2022, from https://healthembassy.co.uk/en/bark/39-cramp-bark.html

Cranberry. (2020, May). National Center for Complementary and Integrative
Health. https://www.nccih.nih.gov/health/cranberry

Cranberry. (2021). In Encyclopædia Britannica. https://www.britannica.-
com/plant/cranberry

Cronkleton, E. (2019, March 8). *10 benefits of lemon balm and how to use it.*
Healthline. https://www.healthline.com/health/lemon-balm-uses

*Culturally and economically important nontimber forest products of northern Maine
- Sustaining forests.* (2010, May 24). United States Department of Agriculture: Forest Service - Northern Research Station. https://www.nrs.fs.fed.
us/sustaining_forests/conserve_enhance/special_products

/maine_ntfp/plants/raspberry/#:~:text=
Habitat%3A%20Raspberries%20are%20often%20found

Cunha, J. P. (Ed.). (2021a, March 9). *Valerian.* Rx List. https://www.rxlist.-
com/consumer_valerian/drugs-condition.htm

Cunha, J. P. (2021b, August 24). *Cranberry.* Rx List. https://www.rxlist.-
com/consumer_cranberry/drugs-condi-
tion.htm#:~:text=Class%3A%20Urology%2C%20Herbals-

Dallmeier, L. (n.d.). *How to make macerated oils. Formula Botanica.* Retrieved
February 24, 2022, from https://formulabotanica.com/how-to-make-
macerated-
oils/#:~:text=Macerating%20works%20best%20when%20done

Dandelion. (n.d.). Mount Sinai Health System. Retrieved February 16, 2022,
from https://www.mountsinai.org/health-library/herb/dande-
lion#:~:text=The%20leaves%20are%20used%20to

Dandelions: Cheery signs of spring. (n.d.). Christian Science Monitor. Retrieved
February 16, 2022, from https://www.csmonitor.com/The-Culture/The-
Home-Forum/2008/0410/p19s01-hfes.html

Davidson, K. (2020, August 20). *Red clover: Benefits, uses, and side effects.*
Healthline. https://www.healthline.com/nutrition/red-clover#benefits

Debret, C. (2021, July 20). *10 useful tools for foraging this summer.* One Green
Planet. https://www.onegreenplanet.org/lifestyle/useful-tools-for-forag-
ing-this-summer/

Dellwo, A. (2020, June 14). *What is Goldenrod?* Very Well Health.
https://www.verywellhealth.com/goldenrod-benefits-4586964

Dessinger, H. (n.d.). *Plantain herb benefits, recipes & how to identify.* Mommypotamus. Retrieved March 2, 2022, from https://mommypotamus.com/plantain/

Dipsacus fullonum. (n.d.). Plants for a Future. Retrieved March 4, 2022, from https://pfaf.org/user/Plant.aspx?LatinName=Dipsacus+fullonum#:~:text=Medicinal%20Uses&text=Teasel%20is%20little%20used%20in

Dodrill, T. (n.d.-a). *How to identify chicory.* New Life on a Homestead. Retrieved February 12, 2022, from https://www.newlifeonahomestead.com/how-to-identify-chicory/#:~:text=The%20basal%20leaves%20on%20the

Dodrill, T. (n.d.-b). *How to identify goldenrod (plus foraging tips).* New Life on a Homestead. Retrieved February 24, 2022, from https://www.newlifeonahomestead.com/goldenrod/#:~:text=Goldenrod%20-plant%20leaves%20have%20only

Douglas, J. (2021, May 6). *Ethical foraging–Responsibility and reciprocity.* Organic Growers School. https://organicgrowersschool.org/ethical-foraging-responsibility-and-reciprocity/#:~:text=Ethical%20foraging%20is%20an%20ongoing

Duiker, S. W., & Curran, W. C. (2007, October 30). *Management of red clover as a cover crop.* Penn State Extension. https://extension.psu.edu/management-of-red-clover-as-a-cover-crop#:~:text=Red%20clover%20does%20best%20on

Dyer, M. H. (2021, August 12). *Mullein herb plants – Tips on using mullein as herbal treatments.* Gardening Know-How. https://www.gardening-knowhow.com/ornamental/flowers/mullein/using-mullein-as-herbs.htm

Echinacea. (n.d.). In The Gale Encyclopedia of Diets: A Guide to Health and Nutrition. Cengage; Encyclopedia. Retrieved February 22, 2022, from https://www.encyclopedia.com/plants-and-animals/plants/plants/echinacea#:~:text=the%20United%20States.-

Echinacea information. (n.d.). Mount Sinai Health System. Retrieved February 22, 2022, from https://www.mountsinai.org/health-library/herb/echinacea

Echinacea purpurea (Eastern purple coneflower). (n.d.). Minnesota Wildflowers. Retrieved February 22, 2022, from https://www.minnesotawildflowers.info/flower/eastern-purple-coneflower#:~:text=Flowers%20are%20single%20on%20end

Echinacea purpurea - Purple cone flower - Medicinal perennial herbal / Flower - 100 seeds. (n.d.). Seeds for Africa. Retrieved February 22, 2022, from https://www.seedsforafrica.co.za/products/echinacea-purpurea-purple-cone-flower-medicinal-annual-flower-100-seeds#:~:text=It%20is%20native%20to%20eastern

Agrimony. (2019, August 14). Britannica. https://www.britannica.com/plant/agrimony

Elder tree remedies. (2015, May 14). Handmade Apothecary.

https://www.handmadeapothecary.co.uk/blog/2015/5/12/elderflow-erpower

Elderberry. (2020, September 21). Web MD. https://www.webmd.com/diet/elderberry-health-benefits#:~:text=The%20berries%20and%20flowers%20of

Elderberry cold and cough syrup recipe. (n.d.). Edible Wild Food. Retrieved February 22, 2022, from https://www.ediblewildfood.com/elderberry-cold-cough-syrup.aspx

Elderberry: Sambucus canadensis. (n.d.). Edible Wild Food. Retrieved February 22, 2022, from https://www.ediblewildfood.com/elderberry.aspx#:~:text=Elder%20is%20characterised%20by%20its

Elderflower. (2021, November 6). Rx List. https://www.rxlist.com/elderflower/supplements.htm

Elderflower tincture - Easy homemade recipe. (2021, June 3). Practical Frugality. https://www.practicalfrugality.com/elderflower-tincture-recipe/

Ellen. (2016, December 16). *Wild garlic (aka field garlic, aka allium vineale).* Backyard Forager. https://backyardforager.com/wild-garlic-field-garlic-allium-vineale/#:~:text=Wild%20garlic%20flowers%20are%20edible

Ellis, N. (2021, April 5). *How to make sunflower oil in your homestead?* Farm & Animals. https://farmandanimals.com/how-to-make-sunflower-oil/

Erdemir, S. M. (2017, September 21). *Parts of a composite flower.* Garden Guides. https://www.gardenguides.com/123630-parts-composite-flower.html

Ripe for the Picking: Blackberry Harvesting Tips and Recipes. (2018, July 17). Espoma Organic. https://www.espoma.com/fruits-vegetables/ripe-for-the-picking-blackberry-harvesting-tips-and-recipes/

Everything mint. (n.d.). Lifestyle Home Garden. Retrieved February 28, 2022, from https://lifestyle.co.za/mint/#:~:text=Mint%20is%20a%20hardy%2C%20highly

Everything you need to know about everbearing mulberry trees. (2020, August 23). This Old House. https://www.thisoldhouse.com/gardening/21336910/everbearing-mulberry-trees

Facts about garlic mustard. (n.d.). Health Benefits Times. https://www.healthbenefitstimes.com/garlic-mustard/

False Solomon seal: Maianthemum racemosum. (n.d.). Edible Wild Food. Retrieved March 3, 2022, from https://www.ediblewildfood.com/false-solomon-seal.aspx#:~:text=Solomon

Feverfew. (2008, May 31). Wildflower Finder. https://wildflowerfinder.org.uk/Flowers/F/Feverfew/Feverfew.htm#:~:text=Distinguishing%20Feature%20%3A%20The%20few%20white

Feverfew (Tanacetum parthenium). (n.d.). Illinois Wildflowers. Retrieved February 23, 2022, from https://www.illinoiswildflowers.info/weeds/plants/feverfew.html#:~:text=

Feverfew%20(Tanacetum%20parthenium)&text=
Description%3A%20This%20perennial%20herbaceous%20plant

Feverfew growing guide. (n.d.). Grow Veg. Retrieved February 23, 2022, from https://www.growveg.co.za/plants/south-africa/how-to-grow-feverfew/

Fewell, A. K. (2020, September 2). *Medicinal uses of goldenrod & goldenrod tincture.* Amy K. Fewell: The Fewell Homestead. https://thefewellhomestead.com/medicinal-uses-of-goldenrod-goldenrod-tincture/

Fletcher, J. (2019, January 3). *Dandelion: Health benefits, research, and side effects.* Medical News Today. https://www.medicalnewstoday.com/articles/324083

Foraging and using birch: Bark, leaves, & sap. (2020, December 1). Grow Forage Cook Ferment. https://www.growforagecookferment.com/foraging-birch/

Ford, C. (2019, February 15). *Cramp bark. Ford's Herbal & Doula Services.* https://www.fordsherbaldoulaservices.com/fords-herb-diary/cramp-bark

Forest Health Staff. (n.d.). *Common mullein - Verbascum thapsus L.* USDA Forest Service. Retrieved March 2, 2022, from http://www.na.fs.fed.us/fhp/invasive_plants

Foster, J. (n.d.). *The easiest way to harvest echinacea seeds.* Grow It Build It. Retrieved February 22, 2022, from https://growitbuildit.com/harvest-echinacea-seeds-an-illustrated-guide/#:~:text=A%20couple%20of%20weeks%20after

Fragrant trees. (n.d.). Berkshire Natural Resources Council. Retrieved February 11, 2022, from https://www.bnrc.org/fragrant-trees/

Frey, M. (2021a, October 12). *The health benefits of echinacea: Can a tea made from purple cone flowers stave off colds and illnesses?* Very Well Fit. https://www.verywellfit.com/echinacea-tea-benefits-and-side-effects-4163612#toc-dosage-and-preparations

Frey, M. (2021b, October 18). *Comfrey tea benefits and side effects is comfrey root safe or healthy?* Very Well Fit. https://www.verywellfit.com/comfrey-tea-benefits-and-side-effects-4163901

Frey, M. (2021c, October 19). *The health benefits of linden: The flowers of this herb are said to have sedative powers.* Very Well Fit. https://www.verywellfit.com/linden-tea-benefits-and-side-effects-4163720

Garlic mustard. (n.d.). Ontario's Invading Species Awareness Program. Retrieved February 24, 2022, from http://www.invadingspecies.com/invaders/plants/garlic-mustard/#:~:text=How%20to%20Identify%20Garlic%20Mustard

Garlic Mustard – A foraging guide to its food, medicine and other uses. (n.d.). Eat Weeds. Retrieved February 24, 2022, from https://www.eatweeds.co.uk/garlic-mustard-alliaria-petiolata#:~:text=Garlic%20mustard%20has%20been%20used,feet%20to%20relieve%20the%20cramp.

Garlic mustard: A very nutritious invasive plant. (2019, January 27). Freak of Natural. https://freakofnatural.com/garlic-mustard/#:~:text=In%20tradi-tional%20herbalism%20garlic%20mustard

German chamomile Information. (n.d.). Mount Sinai Health System. Retrieved February 24, 2022, from https://www.mountsinai.org/health-library/herb/german-chamomile#:~:text=Animal%20stud-ies%20have%20shown%20that

Gerow. (2019, October 15). *Stop and smell the blackberries (then, kill them): A mountain biker's guide to nature.* Singletracks Mountain Bike News. https://www.singletracks.com/mtb-trails/stop-and-smell-the-blackber-ries-a-mountain-bikers-guide-to-nature/

Ghoshal, M. (2020, March 13). *Mulling over mullein leaf.* Healthline. https://www.healthline.com/health/mullein-leaf#mullein-oil

Girvin, T. (2010, August 20). *The scent of birch tar.* Girvin. https://www.girvin.com/the-scent-of-birch-tar/

Goldenrod. (2021, June 11). Rx List. https://www.rxlist.com/goldenrod/supplements.htm

Goldenrod: Medicinal uses & benefits. (n.d.). Chestnut School of Herbal Medi-cine. Retrieved February 24, 2022, from https://chestnutherbs.com/medi-cinal-uses-and-benefits-of-goldenrod/

Gotter, A. (2018, September 18). *Catnip tea: Health benefits and uses.* Healthline. https://www.healthline.com/health/catnip-tea#side-effects-and-risks

Gotter, A. (2021, January 12). *What is burdock root?* Healthline. https://www.healthline.com/health/burdock-root

Grant, A. (2021, August 8). *Borage harvesting: How and when to harvest borage plants.* Gardening Know-How. https://www.gardeningknowhow.com/ed-ible/herbs/borage/harvesting-borage-plants.htm

Grant, B. L. (2020, December 20). *Leaf identification—Learn about different leaf types in plants.* Gardening Know-How. https://www.gardeningknowhow.com/garden-how-to/info/different-leaf-types-in-plants.htm

Griffin, R. M. (2021, January 18). *Black cherry and your health.* Web MD. https://www.webmd.com/diet/supplement-guide-black-cher-ry#:~:text=Black%20cherry%20bark%20also%20seems

Growing & Foraging for Mullein (Plus Harvesting & Preserving tips!). (2021, June 10). Unruly Gardening. https://unrulygardening.com/growing-foraging-mullein/#:~:text=Summer%2C%20or%20whenever%20the%20plant

Growing Organic Mugwort from Seed to Harvest. (n.d.). Mary's Heirloom Seeds. Retrieved February 28, 2022, from https://www.marysheirloomseeds.com/blogs/news/78072001-growing-organic-mugwort-from-seed-to-harvest#:~:text=Harvest%20mugwort%20shortly%20before%20it

Haddock, B. (2012, May 11). *Cranberry - edible wild plant - how to find, identify, prepare, and other uses for survival.* Wilderness Arena. https://www.wildernessarena.com/food-water-shelter/food-food-water-shelter/food-procurement/edible-wild-plants/cranberry

Haines, A. (n.d.). *Why foraging?* Arthur Haines. http://www.arthurhaines.com/why-foraging#:~:text=It%20helps%20people%20become%20more

Hall, J. (2021, July 9). *How to find and prepare nutritious, edible mallows.* Den Garden.
https://dengarden.com/gardening/malva#:~:text=Identifying%20Mallows

Hanrahan, C., & Frey, R. (2018). *Mugwort.* In Gale Encyclopedia of Alternative Medicine. Cengage; Encyclopedia. https://www.encyclopedia.com/medicine/drugs/pharmacology/mugwort#:~:text=
Mugwort%20is%20a%20tall%20and

Harrington, J. (n.d.). *How to harvest echinacea for tea.* SF Gate. Retrieved February 22, 2022, from https://homeguides.sfgate.com/harvest-echinacea-tea-73456.html

Harris, J. (n.d.). *My homemade echinacea tincture.* Jillian Harris. Retrieved February 22, 2022, from https://jillianharris.com/

Harvesting birch bark. (n.d.). The Folk School Fairbanks. Retrieved February 11, 2022, from https://folk.school/classes/tutorials/harvesting-birch-bark/

Hatter, K. (2017, July 21). *How to identify camomile.* Garden Guides. https://www.gardenguides.com/13426932-how-to-identify-camomile.html

Health benefits of mullein tea. (n.d.). Web MD. Retrieved March 2, 2022, from https://www.webmd.com/diet/health-benefits-mullein-tea#1

Health benefits of smooth Solomon's seal. (2018, November 5). Health Benefits Times. https://www.healthbenefitstimes.com/smooth-solomons-seal/

Health benefits of valerian root. (n.d.). Web MD. Retrieved March 4, 2022, from https://www.webmd.com/diet/health-benefits-valerian-root#3

Heath, S. (2021, December 9). *How to grow a mullein plant.* The Spruce. https://www.thespruce.com/mullein-plant-growing-guide-5203326

Hegde, R. (2020, August 12). *How to make willow bark tea for pain relief.* Organic Facts. https://www.organicfacts.net/willow-bark-tea.html

Herb spotlight - Cramp bark. (n.d.). Sun God Medicinals. Retrieved February 15, 2022, from https://sungodmedicinals.com/pages/herb-spotlight-cramp-bark

Herb: Japanese Honeysuckle. (n.d.). Natural Medicinal Herbs. Retrieved February 24, 2022, from http://www.naturalmedicinalherbs.net/herbs/l/lonicera-japonica=japanese-honeysuckle.php#:~:text=Edible%20parts%20of%20Japanese%20Honeysuckle%3A&text=
The%20parboiled%20leaves%20are%20used

Herbal, O. W. (2014, November 4). *Bark harvest & ethical wildcrafting.* Old Ways Herbal: Juliette Abigail Carr. https://oldwaysherbal.com/2014/11/04/ethical-wildcrafting/

Hodgson, D. (n.d.). *Cleavers ointment.* Woodland Ways. Retrieved February 12, 2022, from https://www.woodland-ways.co.uk/blog/hedgerow-medi-cines/cleavers-ointment/

Homemade blackberry tincture. (2021, August 31). Alco Reviews. https://alcore-views.com/homemade-blackberry-tincture/

Honeysuckle. (2021, November 6). Rx List. https://www.rxlist.com/honey-suckle/supplements.htm

Horsetail. (2014, June 5). Britannica. https://www.britannica.com/plant/horsetail

Horsetail. (2021, June 11). Rx List. https://www.rxlist.com/horsetail/supple-ments.htm

How cranberries grow: Pollination. (n.d.). Massachusetts Cranberries. Retrieved February 16, 2022, from https://www.cranberries.org/how-cranberries-grow/pollination

How much cherry juice should you drink a day? Traverse Bay Farms. www.tra-versebayfarms.com/pages/cherries-recommended-dosage-of-cherry-juice. Accessed 16 Mar. 2022.

How to do a coffee enema. (2020, November 9). Pure Joy Planet. https://www.purejoyplanet.com/blog/how-to-do-a-coffee-enema

How to grow & use valerian. (2020, June 24). It's My Sustainable Life. https://www.itsmysustainablelife.com/how-to-grow-use-valer-ian/#:~:text=Ideally%2C%20valerian%20root%20should%20not

How to grow big, tall sunflowers. (2019, June 1). Velcro Brand Blog. https://www.velcro.com/news-and-blog/2019/06/how-to-grow-big-tall-sunflowers/#:~:text=Sunflowers%20prefer%20a%20somewhat%20alka-line,before%20you%20do%20any%20planting!

How to harvest wild cherry bark and stop coughing so you can sleep. (n.d.). Joybilee Farm. Retrieved February 11, 2022, from https://joybileefarm.com/wild-cherry-bark-stop-coughing/

How to identify chickweed - Foraging for wild edible greens. (2017, December 15). Good Life Revival. https://thegoodliferevival.com/blog/chick-weed#:~:text=Chickweed%20is%20easy%20to%20identi-fy,each%20other%20along%20the%20stem.

How to identify elderflower. (2012, June 8). Stay & Roam. https://stayan-droam.blog/how-to-identify-elderflower/#:~:text=the%20correct%20plant.-

How to make birch bark flour and bake with it. (2020, December 18). Tree Time. https://blog.treetime.ca/blog/how-to-make-and-bake-with-bark-flour/

How to make Dandelion Tea the perfect way each time!! (n.d.). Tea Swan. Retrieved February 16, 2022, from https://teaswan.-com/blogs/news/how-to-make-dandelion-tea

How to make echinacea (purple coneflower) oil & salve. (n.d.). The Nerdy Farm

Wife. Retrieved February 22, 2022, from https://thenerdyfarmwife.-com/echinacea-purple-coneflower-oil-salve/

How to Make Homemade Echinacea Tea. (2022, October 2). Sencha Tea Bar. https://senchateabar.com/blogs/blog/how-to-make-echinacea-tea#:~:text=Echinacea%20tea%20can%20be%20made

Howland, G. (2018 6). *Red raspberry leaf tea recipes you'll actually want to drink.* Mama Natural. https://www.mamanatural.com/red-raspberry-leaf-tea-recipes/

Iannotti, M. (2021a, July 9). *How to grow borage.* The Spruce. https://www.thespruce.com/how-to-grow-borage-1402625#toc-harvest-ing-borage

Iannotti, M. (2021b, December 20). *How to grow and care for chamomile.* The Spruce. https://www.thespruce.com/how-to-grow-chamomile-1402627

Identify a mulberry - Find out what to look out for to hunt down a mulberry. (n.d.). Morus Londinium. Retrieved March 1, 2022, from https://www.morus-londinium.org/map/identify

Identifying burdock. (2014, October 20). The Druid's Garden. https://druidgar-den.wordpress.com/tag/identifying-burdock/

International Culinary Center. (2018, November 28). *How to Safely Forage.* International Culinary Education. https://www.ice.edu/blog/how-to-safely-forage

J, C. (2018, December 21). *A refreshing, aromatic herbal aperitif.* The Inspired Home. https://theinspiredhome.com/articles/a-refreshing-aromatic-herbal-aperitif

Jaana. (n.d.). *Larch.* Herbal Picnic. Retrieved February 28, 2022, from https://herbalpicnic.blogspot.com/2015/04/larch.html

James. (2020a, April 29). *Burdock (arctium iappa) identification.* Totally Wild UK. https://totallywilduk.co.uk/2020/04/29/identify-burdock/#:~:text=Burdock%20will%20grow%20in%20pretty

James. (2020b, April 29). *Cleavers (gallium aparine) identification.* Totally Wild UK. https://totallywilduk.co.uk/2020/04/29/identify-cleavers/

James. (2020c, April 29). *Comfrey (symphytum officinale) identification.* Totally Wild UK. https://totallywilduk.co.uk/2020/04/29/identify-comfrey/

James. (2020d, April 29). *Yarrow (achillea millefolium) identification.* Totally Wild UK. https://totallywilduk.co.uk/2020/04/29/identify-yarrow/

James, T. (2016, January 19). *How to forage for wild catnip.* Adventure Cats. https://www.adventurecats.org/pawsome-reads/foraging-adventure-how-to-spot-wild-catnip/#:~:text=Catnip%20is%20grayish%2Dgreen%20and

Japanese honeysuckle (lonicera japonica). (n.d.). Invasive. Retrieved February 28, 2022, from https://www.invasive.org/alien/pubs/midatlantic/lo-ja.htm#:~:text=It%20is%20a%20fast%2Dgrowing

Jeanroy, A. (2019, June 26). *How to make herbal infusions.* The Spruce Eats.

https://www.thespruceeats.com/how-to-make-an-herbal-infusion-1762142

Joanna. (2014, July 9). *The benefits of raspberry leaf tincture & my raspberry leaf tincture recipe.* Joanna Steven. https://www.joannasteven.com/the-benefits-of-raspberry-leaf-tincture-my-raspberry-leaf-tincture-recipe/

Johnson, K. (2000, August). *Rubus ursinus.* Pdx. http://web.pdx.edu/~maserj/ESR410/rubisursinus.html

Jones, L. (2017, October 11). *Homegrown blackberry leaf tea.* Windellama Organics. https://www.windellamaorganics.com/blog/2017/10/10/homegrown-blackberry-leaf-tea

Keeler, K. (n.d.). *Plant story: Common mullein and its folklore.* A Wandering Botanist. Retrieved March 2, 2022, from http://khkeeler.blogspot.com/2013/07/plant-story-common-mullein-and-its.html#:~:text=Histori cally%20mullein%20was

%20considered%20a,to%20be%20their%20preferred

%20torch.&text=Common%20mullein

%20has%20yellow%20flowers,Mercury

%20gave%20Odysseus%20common%20mullein.

Kendle. (2018, September 14). *How to make herbal infusions & decoctions for wellness support.* Mountain Rose Herbs. https://blog.mountainroseherbs.com/herbal-infusions-and-decoctions

Kirk, S., Belt, S., & Berg, N. A. (2011). *Plant fact sheet verbena hastata (L.) Plant symbol = VEHA2.* Natural Resources Conservation Service. National Plant Materials Center. https://www.nrcs.usda.gov/Internet/FSE_PLANTMA-TERIALS/publications/mdpmcfs10335.pdf

Krohn, E. (2016). Alder. *Wild Foods and Medicines.* http://wildfoodsandmedicines.com/alder/

Kubala, J. (2019, November 29). *Mulberry leaf: Uses, benefits, and precautions.* Healthline. https://www.healthline.com/nutrition/mulberry-leaf#benefits

Lang, A. (2020, June 8). *What is vervain? All you need to know.* Healthline. https://www.healthline.com/nutrition/vervain-verbena

Lapcevic, K. (2021, July 21). *Lemon balm tincture.* Homespun Seasonal Living. https://homespunseasonalliving.com/lemon-balm-tincture/

Larch. (n.d.-a). Muster Kiste. Retrieved February 24, 2022, from http://www.musterkiste.com/en/holz/pro/1016_Larch.html#:~:text= The%20larch%20is%20the%20heaviest

Larch. (n.d.-b). Gaia Herbs. Retrieved February 28, 2022, from https://www.gaiaherbs.com/blogs/herbs/larch

Larch. (n.d.-c). Medicinal Herb Info. Retrieved February 28, 2022, from http://medicinalherbinfo.org/000Herbs2016/1herbs/larch/

Larch. (n.d.-d). Alaska's Wilderness Medicines. Retrieved February 28, 2022, from http://www.ankn.uaf.edu/curriculum/Books/Viereck/viereck-larch.html

Larch. (n.d.-e). Dr. Hauschka. Retrieved February 28, 2022, from https://www.drhauschka.co.uk/medicinal-plant-glossary/larch/

Larch arabinogalactan: Uses, side effects, and more. (2019). Web MD. https://www.webmd.com/vitamins/ai/ingredientmono-974/larch-arabinogalactan

Lemm, E. (2021, July 23). *Wild garlic adds subtle flavor.* The Spruce Eats. https://www.thespruceeats.com/what-is-wild-garlic-435437

Lemon balm. (n.d.). Mount Sinai Health System. Retrieved February 28, 2022, from https://www.mountsinai.org/health-library/herb/lemon-balm#:~:text=Lemon%20balm%20(Melissa%20officinalis)%2C

Lemon balm (melissa officinalis). (n.d.). Illinois Wildflowers. Retrieved February 28, 2022, from https://www.illinoiswildflowers.info/weeds/plants/lemon_balm.html

Lemon balm – Identification, edibility, distribution, ecology. (2019, July 9). Galloway Wild Foods. https://gallowaywildfoods.com/lemon-balm-identification-edibility-distribution-ecology/#:~:text=Identification%20%E2%80%93%204%2F5%20%E2%80%93%20Lemon

Lemon balm cream. (2021, April 7). Herba Zest. https://www.herbazest.com/herbs/lemon-balm/lemon-balm-cream

Lemon balm production (Second). (2012). Department of Agriculture, Forestry and Fisheries | Directorate: Pant production. chrome-extension://efaidnbmnnnibpcajpcglclefindmkaj/viewer.html?pdfurl=https%3A%2F%2Fwww.dalrrd.gov.za%2FPortals%2F0%2FBrochures%2520and%2520Production%2520guidelines%2FProduction%2520Guidelines%2520Lemon%2520Balm.pdf&clen=1437819&chunk=true

Lime tree in culture. (n.d.). Wikipedia. Retrieved February 28, 2022, from https://en.wikipedia.org/wiki/Lime_tree_in_culture#:~:text=Germanic%20mythology

Linden. (n.d.). Drugs. Retrieved February 28, 2022, from https://www.drugs.com/npc/linden.html#:~:text=Linden%20has%20been%20used%20to

Linden. (2021, June 11). Rx List. https://www.rxlist.com/linden/supplements.htm

Linden trees: Types, leaves, flowers, bark – Identification (with pictures). (n.d.). Leafy Place. Retrieved February 28, 2022, from https://leafyplace.com/linden-trees/#:~:text=size%20and%20shape.-

Link, R. (2020, June 10). *What is plantain weed, and how do you use it?* Healthline. https://www.healthline.com/nutrition/plantain-weed#side-effects

Lisa, A. (2022, January 18). *How to extract oil from plants (plus the numerous benefits and uses).* The Practical Planter. https://thepracticalplanter.com/how-to-extract-oil-from-plants/

Lofgren, K. (2020, May 27). *What's the difference between English and German chamomile?* Gardener's Path.

https://gardenerspath.com/plants/flowers/english-german-chamomile/#German-Chamomile

Longacre, C. (2020, September 12). *How to harvest sunflower seeds.* Almanac. https://www.almanac.com/harvesting-sunflower-seeds

Making birch bark oil. (2018, January 24). Plant Pioneers. https://www.plant-pioneers.org/blog/2018/1/24/making-birch-bark-oil

Making catnip (nepeta cataria) tincture. (2014, September 2). Spiraea. https://spiraeaherbs.ca/making-catnip-nepeta-cataria-tincture/

Mallow. (2015, May 28). Peace Health. https://www.peacehealth.org/medical-topics/id/hn-3263004#hn-3263004-how-it-works

Mallow. (n.d.). E Medicine Health. Retrieved February 13, 2022, from https://www.emedicinehealth.com/mallow/vitamins-supplements.htm#UsesAndEffectiveness

Mallow - common (malva sylvestris) organically grown flower seeds. (n.d.). Floral Encounters. Retrieved February 13, 2022, from https://www.floralen-counters.com/Seeds/seed_detail.jsp?grow=Mallow+-+Common&pro-ductid=1102

Mallow is medicine. (2018, September 29). Indigenous Goddess Gang. https://www.indigenousgoddessgang.com/self-care-medi-cine/2018/9/27/white-flowered-medicine

Mallow: Overview, uses, side effects, and more. (n.d.). Web MD. Retrieved February 12, 2022, from https://www.webmd.com/vitamins/ai/ingredi-entmono-192/mallow

Matricaria chamomilla (wild chamomile). (n.d.). Go Botany: Native Plant Trust. Retrieved February 24, 2022, from https://gobotany.nativeplant-trust.org/species/matricaria/chamomilla/

Matthews, C. (Ed.). (2022, January 12). *Wild garlic guide: Where to find, how to cook it and recipe ideas.* Countryfile. https://www.countryfile.com/how-to/food-recipes/wild-garlic-guide-where-to-find-how-to-cook-it-and-recipe-ideas/#:~:text=What%20are%20the%20health%20benefits,as%20heart%20attack%20or%20stroke.

May, D. (2011, September 6). *Nature notes: common mallow.* The Times. https://www.thetimes.co.uk/article/nature-notes-common-mallow-lmrrwlmh76x

McCulloch, M. (2018, November 22). *Are sunflower seeds good for you?* Nutrition, benefits and more. Healthline. https://www.healthline.com/nutri-tion/sunflower-seeds#eating-tips

McCulloch, M. (2019, April 4). *Goldenrod: Benefits, dosage, and precautions.* Healthline. https://www.healthline.com/nutrition/goldenrod#inflammation

McKenzie, R. (2019). *Solomon's seal.* Eclectic School of Herbal Medicine. https://www.eclecticschoolofherbalmedicine.com/solomons-seal/

McMinn, S. (n.d.). *How to steep sassafras roots.* Chickens in the Road. Retrieved

March 3, 2022, from https://chickensintheroad.com/cooking/how-to-steep-sassafras-roots/#:~:text=The%20best%20time%20to%20dig

Medicinal uses of beech trees. (n.d.). What Tree Where. Retrieved February 11, 2022, from https://whattreewhere.com/tag/medicinal-uses-of-beech-trees/

Melissa officinalis L. (n.d.). A Vogel. Retrieved February 28, 2022, from https://www.avogel.com/plant-encyclopaedia/melissa_officinalis.php#:~:text=Lemon%20Balm%20is%20a%2050cm

Mentha spicata - Spearmint. (n.d.). Northwest Oregon Wetland Plants Project. Retrieved February 28, 2022, from http://web.pdx.edu/~maserj/ES-R410/mentha.html#:~:text=Habitat

Meredith, L. (2021, September 14). *Violet flower syrup.* The Spruce Eats. https://www.thespruceeats.com/violet-flower-syrup-recipe-1327872

Merva, V. (2020, March 20). *How to make Violet oil and its uses.* Simply beyond Herbs. https://simplybeyondherbs.com/violet-oil-recipe/

Minifie, K. (2014, June 5). *What the heck is chickweed and why is it on my plate?* Epicurious. https://www.epicurious.com/archive/blogs/editor/2014/05/what-the-heck-is-chickweed-and-why-is-it-on-my-plate.html

Mint. (2021, July 5). Encyclopedia Britannica. https://www.britannica.com/plant/Mentha

Moline, P. (2015, June 26). *How to make your own herb tincture or peppermint oil.* Los Angeles Times. https://www.latimes.com/home/la-he-healing-garden-recipes-20150627-story.html

Monoecious. (n.d.). In Merriam Webster. Retrieved February 8, 2022, from https://www.merriam-webster.com/dictionary/monoecious

Mugwort. (n.d.). Sawmill Herb Farm. Retrieved February 28, 2022, from http://www.sawmillherbfarm.com/herb%20profile/mugwort/

Mugwort. (2021, June 11). Rx List. https://www.rxlist.com/mugwort/supplements.htm#SpecialPrecautionsWarnings

Mulberry tree facts that are absolutely compelling to read. (n.d.). Gardenerdy. Retrieved March 1, 2022, from https://gardenerdy.com/mulberry-tree-facts/

Nash, K. (2021, June 5). *Where do pine trees grow? (Best habitat for natural growth).* Tree Vitalize. https://www.treevitalize.net/where-do-pine-trees-grow/#:~:text=agriculture%2C%20or%20fire.-

Nielsen, L. (2021, January 5). *How to harvest mint and store it for later.* Epic Gardening. https://www.epicgardening.com/how-to-harvest-mint/#:~:text=When%20Should%20I%20Harvest%20Mint%20Leaves%3F

Nix, S. (2021a, May 7). *Identify the larch.* Treehugger. https://www.treehugger.com/identify-the-larch-1341861#:~:text=How%20to%20Identify%20Larches&text=Most%20common%20larches%20in%20North

Nix, S. (2021b, June 7). *How to identify a tree using leaf shape, margin, and vena-*

tion. Treehugger. https://www.treehugger.com/id-trees-using-leaf-shape-venation-1343511

Types of Oak trees with their bark and leaves – Identification guide (pictures). (n.d.). Leafy Place. Retrieved March 2, 2022, from https://leafyplace.com/oak-tree-types-bark-leaves/#:~:text=The%20bark%20of%20young%20oak

Oak. (2020, June 18). Herba Zest. https://www.herbazest.com/herbs/oak

Oak - A foraging guide to its food, medicine and other uses. (n.d.). Eat Weeds. Retrieved March 2, 2022, from https://www.eatweeds.co.uk/oak-quer-cus-robur#:~:text=sugars%20and%20tannins.-

Oak bark. (2021, June 11). Rx List. https://www.rxlist.com/oak_bark/supple-ments.htm#SpecialPrecautionsWarnings

Oak flowers. (n.d.). Backyard Nature. Retrieved March 2, 2022, from https://www.backyardnature.net/fl_bloak.htm

Oak mythology and folklore. (n.d.). Trees for Life. Retrieved March 2, 2022, from https://treesforlife.org.uk/into-the-forest/trees-plants-animals/trees/oak/oak-mythology-and-folklore/#:~:text=To%20the%20-Greeks%2C%20Romans%2C%20Celts

Oak Tree. (n.d.). Tree Works: Qualified Tree Surgery. Retrieved March 2, 2022, from http://treeworksguernsey.co.uk/tree-identification/oak/

Oak tree facts. (n.d.). Soft Schools. Retrieved March 2, 2022, from https://www.softschools.com/facts/plants/oak_tree_facts/505/

Orr, E. (n.d.). *Stinging nettle: Where to find & how to identify.* Wild Edible. Retrieved March 3, 2022, from https://www.wildedible.com/wild-food-guide/stinging-nettle#:~:text=Nettles%20grow%202%20to%205

Parallel venation. (2021, June 17). Maximum Yield. https://www.maxi-mumyield.com/definition/1733/parallel-venation

Parisian, K. (2020, December 17). *Infused oils benefits and warnings.* Parisian's Pure Indulgence. https://parisianspure.com/blogs/news/infused-oils-benefits-and-warnings

Pearson, K. (2017, December 13). *8 health benefits of mint.* Healthline. https://www.healthline.com/nutrition/mint-benefits#TOC_TI-TLE_HDR_10

Perforate St John's-wort. (n.d.). The Wildlife Trusts. Retrieved March 3, 2022, from https://www.wildlifetrusts.org/wildlife-explorer/wildflowers/per-forate-st-johns-wort#:~:text=Perforate%20st%20John

Phillips, Q. (2019, September 4). *How to make valerian tea.* Everyday Health. https://www.everydayhealth.com/diet-nutrition/make-valerian-tea-how-prepare-brew-steep-this-herbal-tea/

Pietrangelo, A. (2018, June 7). *St. John's wort: The benefits and the dangers.* Healthline. https://www.healthline.com/health-news/is-st-johns-wort-safe-080615

Pillsbury, C. (2017, August 23). *Forage through borage seed oil.* Watson. https://blog.watson-inc.com/nutri-knowledge/forage-through-borage-seed-oil

Pine. (2021, June 11). Rx List. https://www.rxlist.com/pine/supplements.htm#UsesAndEffectiveness

Plantain. (2015, May 24). Kaiser Permanente. https://wa.kaiserpermanente.org/kbase/topic.jhtml?docId=hn-2148003

Poulson, B., Horowitz, D., & Trevino, H. M. (Eds.). (n.d.). *Feverfew.* University of Rochester Medical Center. Retrieved February 23, 2022, from https://www.urmc.rochester.edu/encyclopedia/content.aspx?contenttypeid=19&contentid=Feverfew#:~:text=Feverfew%20may%20reduce%20painful%20inflammation

Product information teaseltea. (n.d.). Teasel Shop. Retrieved March 4, 2022, from https://www.teaselshop.com/c-2283855/dosage-and-use/

Pruisis, E. (2021, August 9). *Discover varieties of alder trees and shrubs.* The Spruce. https://www.thespruce.com/alder-trees-and-shrubs-3269701

Raman, R. (2018, October 25). *Echinacea: Benefits, uses, side effects and dosage.* Healthline. https://www.healthline.com/nutrition/echinacea#:~:text=The%20bottom%20line-

Red Clover. (n.d.). Brandeis University. Retrieved March 3, 2022, from http://www.bio.brandeis.edu/fieldbio/EFG_DEB_SHU/species%20-pages/Red%20Clover/Red%20Clover.html#:~:text=Identifying%20Characteristics%3A%20The%20white%20V

Red clover. (n.d.-a). Bellarmine University. Retrieved March 3, 2022, from https://www.bellarmine.edu/faculty/drobinson/RedClover.asp

Red clover. (n.d.-b). Plant Life. Retrieved March 3, 2022, from https://www.plantlife.org.uk/uk/discover-wild-plants-nature/plant-fungi-species/red-clover#:~:text=Flowers%20May%20to%20September.

Red clover (trifolium pratense). (n.d.). Illinois Wildflowers. Retrieved March 3, 2022, from http://www.illinoiswildflowers.info/weeds/plants/red_clover.htm#:~:text=
Habitats%20include%20fields%2C%
20pastures%2C%20weedy

Red raspberry: Uses, side effects, & more. (n.d.). Web MD. Retrieved March 3, 2022, from https://www.webmd.com/vitamins/ai/ingredientmono-309/red-raspberry

Reeves, K. (2010). *Exotic species: St. Johnswort.* U.S. National Park Service. https://www.nps.gov/articles/st-johnswort.htm#:~:text=Flowers%
20and%20Fruits&text=
along%20the%20margins.-

Rose, S. (n.d.). *Herbal guide to feverfew.* Garden Therapy. Retrieved February 23, 2022, from https://gardentherapy.ca/herbal-guide-to-feverfew/

Rowland, B. (n.d.). *Cramp Bark.* Encyclopedia; Gale Encyclopedia of Alternative Medicine. Retrieved February 15, 2022, from https://www.encyclopedia.com/medicine/encyclopedias-almanacs-transcripts-and-maps/cramp-bark

Sacred tree profile: Cherry (prunus serotina)'s magic, mythology, medicine and meaning. (n.d.). Retrieved February 12, 2022, from https://druidgarden.-wordpress.com/2019/06/23/sacred-tree-profile-cherry-prunus-seroti-nas-magic-mythology-medicine-and-meaning/

Sacred tree profile: Sassafras' medicine, magic, mythology and meaning. (n.d.). The Druid's Garden. Retrieved March 3, 2022, from https://druidgarden.-wordpress.com/2017/08/20/sacred-tree-profile-sassafras-medicine-magic-mythology-and-meaning/

Sarsaparilla. (n.d.). Rx List. Retrieved March 4, 2022, from https://www.rxlist.com/sarsaparilla/supplements.htm

Sassafras. (n.d.). University of Kentucky. Retrieved March 3, 2022, from https://www.uky.edu/hort/Sassafras#:~:text=Introduction%3A%20Sas-safras%20has%20exceptional%20features

Sassafras. (2021, June 11). Rx List. https://www.rxlist.com/sassafras/supple-ments.htm

Sassafras tree: Leaves, flowers, bark (pictures) - Identification guide. (n.d.). Leafy Place. Retrieved March 3, 2022, from https://leafyplace.com/sassafras-tree/#:~:text=Sassafras%20tree%20identification%20is%20by

Schonbeck, J., & Frey, R. (n.d.). *Oak.* In Gale Encyclopedia of Alternative Medicine. Encyclopedia. Retrieved March 2, 2022, from https://www.encyclopedia.com/plants-and-animals/plants/plants/oak#:~:text=oak%20%2F%20C5%8Dk%2F%20E2%80%A2%20n.

Scots pine mythology and folklore. (n.d.). Trees for Life. Retrieved March 2, 2022, from https://treesforlife.org.uk/into-the-forest/trees-plants-animals/trees/scots-pine/scots-pine-mythology-and-folk-lore/#:~:text=Pine%20was%20also%20a%20fertility

Scott, C. (n.d.). *Medicinal plant: Japanese honeysuckle.* George Mason University. Retrieved February 24, 2022, from http://mason.gmu.edu/~cscottm/plants.html

Sedgwick, I. (2019, February 16). *Violets are blue: The folklore of February's birth flower.* Icy Sedgwick. https://www.icysedgwick.com/violets-folklore/

Seed information: Common name: valerian | Scientific name: Valeriana officinalis. (n.d.). Herb Garden. Retrieved March 4, 2022, from https://www.herbgar-den.co.za/mountainherb/seedinfo.php?id=74#:~:text=Seed%20Informa-tion&text=Valerian%20is%20native%20to%20Britain

Seladi-Schulman, J. (2020, June 3). *About wintergreen essential oil.* Healthline. https://www.healthline.com/health/wintergreen-oil#uses

Shepherd's purse. (2021, June 11). Rx List. https://www.rxlist.com/shepherd-s_purse/supplements.htm

Shepherd's purse – and the value of stories. (2020, June 25). Diego Bonetto. https://www.diegobonetto.com/blog/shepherd-purse-and-the-value-of-stories#:~:text=It%20is%20also%20a%20very

Shepherd's purse tincture. (2017, March 13). Women's Heritage. https://www.-womensheritage.com/blog/2017/2/27/shepards-purse-tincture

Shepherds purse: Capsella bursa pastoris. (n.d.). Edible Wild Food. Retrieved March 3, 2022, from https://www.ediblewildfood.com/shepherds-purse.aspx#:~:text=Shepherd

Shore, T. (2018, December 14). *How to identify red raspberry bushes & leaves.* Home Guides. https://homeguides.sfgate.com/identify-red-raspberry-bushes-leaves-56436.html

Silver birch - A foraging guide to its food, medicine and other uses. (n.d.). Eat Weeds. Retrieved February 14, 2022, from https://www.eatweeds.-co.uk/birch-betula-spp#Harvest_Time

Simone. (2015, May 19). *How to make blackberry leaf tincture.* Solar Ripe. http://solarripe.eu/2015/05/how-to-make-blackberry-leaf-tincture-may-2015/

Smith, E. (n.d.). *Plant profile: Violet.* Integrative Family Medicine of Asheville. Retrieved March 4, 2022, from https://www.integrativea-sheville.org/plant-profile-violet/#:~:text=Violet%20is%20moist%20and%20cooling

Snyder, C. (2021, May 14). *What is an herbal tonic? Benefits, weight loss, and efficacy.* Healthline. https://www.healthline.com/nutrition/herbal-tonic

Solomon's seal. (2019, November 18). The Witchipedia. https://witchipedia.-com/book-of-shadows/herblore/solomons-seal/

Solomon's seal. (2021, June 11). Rx List. https://www.rxlist.com/solomon-s_seal/supplements.htm

Spengler, T. (2018, August 23). *How to identify mint plants.* Garden Guides. https://www.gardenguides.com/78357-history-mint-plant.html

Spengler, T. (2020, February 11). *Scale leaf evergreen varieties: What is a scale leaf evergreen tree.* Gardening Know-How. https://www.gardening-knowhow.com/ornamental/trees/tgen/scale-leaf-evergreen-varieties.htm

St. John's Wort. (2020, October). National Center for Complementary and Integrative Health. https://www.nccih.nih.gov/health/st-johns-wort#:~:text=Currently%2C%20St.

Staughton, J. (2020a, July 13). *5 surprising benefits of agrimony tea.* Organic Facts. https://www.organicfacts.net/agrimony-tea.html

Staughton, J. (2020b, July 28). *10 surprising benefits of sassafras.* Organic Facts. https://www.organicfacts.net/health-benefits/other/sassafras.html

Staughton, J. (2020c, August 25). *6 incredible benefits of St John's wort tea.* Organic Facts. https://www.organicfacts.net/st-johns-wort-tea.html

Stein, J., Binion, D., & Acciavatti, R. (2001). *Field guide to native oak species of Eastern North America (FHTET-03-01).* USDA Forest Service, Forest Health Technology Enterprise Team. chrome-extension://efaidnbmnnnibpca jpcglclefindmkaj/viewer.html?pdfurl= https%3A%2F%2Fwww.fs.fed.us%2Fforesthealth%2

Ftechnology%2Fpdfs%2Ffieldguide.pdf&clen=
10803156&chunk=true

Stewart, S. (n.d.). *Mullein tincture.* Just a Pinch Recipes. Retrieved March 2, 2022, from https://www.justapinch.com/recipes/non-edible/non-edible-other-non-edible/mullein-tincture.html

Stinging nettle. (n.d.). Mount Sinai Health System. Retrieved March 3, 2022, from https://www.mountsinai.org/health-library/herb/stinging-nettle#:~:text=Stinging%20nettle%20has%20been%20used

Stobart, A. (2019, April 22). *Making herbal poultices and compresses.* Medicinal Forest Garden Trust. https://medicinalforestgardentrust.org/making-herbal-poultices-and-compresses/#:~:text=Herb%20poul-tices&text=You%20will%20need%20to%20crush

Streit, L. (2019, November 14). *5 emerging benefits and uses of chicory root fiber.* Healthline. https://www.healthline.com/nutrition/chicory-root-fiber#_noHeaderPrefixedContent

Strobile. (n.d.). In Merriam Webster. Retrieved February 8, 2022, from https://www.merriam-webster.com/dictionary/strobile

Susannah. (2021, November 1). *Cleavers plant uses & 5 best cleavers herb benefits.* Healthy Green Savvy. https://www.healthygreensavvy.com/cleavers-plant-herb-benefits/

Sweet violet - Uses, side effects, and more. (n.d.). Web MD. Retrieved March 4, 2022, from https://www.webmd.com/vitamins/ai/ingredientmono-212/sweet-violet#

Tapping tree sap. (2011, March 26). Judy of the Woods. http://www.judyofthe-woods.net/forage/tree_sap.html

Taylor, K. (2021, April 21). *How to dry echinacea flowers And roots.* Urban Garden Gal. https://www.urbangardengal.com/how-to-dry-echinacea-flowers-roots/#:~:text=Echinacea%20roots%20can%20be%20harvested

Tea time: Violet leaf tea. (2020, April 15). Awkward Botany. https://awkward-botany.com/2020/04/15/tea-time-violet-leaf-tea/

Teazle. (2021, June 14). EMedicine Health. https://www.emedicinehealth.-com/teazle/vitamins-supplements.htm

Teazle - Uses, side effects, and more. (n.d.). Web MD. Retrieved March 4, 2022, from https://www.webmd.com/vitamins/ai/ingredientmono-187/teazle

Tello, C. (2021, September 9). What is sarsaparilla? A plant lost in time + tea recipe. Self Decode. https://supplements.selfdecode.com/blog/sarsaparilla-plant-drink-how-to-make-tea/

Terry, S. (2017, September 21). *How to identify willow trees.* Garden Guides. https://www.gardenguides.com/109710-care-corkscrew-willow-tree.html

The benefits of wild cherry bark. (n.d.). Kaya Well. Retrieved February 11, 2022, from https://www.kayawell.com/Food/The-Benefits-of-Wild-Cherry-Bark-whooping-cough-diarrhea

The raspberries are ready for picking. (2019, July 9). The Martha Blog. https://www.themarthablog.com/2019/07/the-raspberries-are-ready-for-picking.html

Tikkanen, A. (n.d.). *Borage.* Encyclopedia Britannica. Retrieved February 12, 2022, from https://www.britannica.com/plant/borage

Tiller, B. (2019, January 24). *Field garlic: Allium vineale.* Mossy Oak. https://www.mossyoak.com/our-obsession/blogs/how-to/field-garlic-allium-vineale

Tilley, N. (2021, July 26). *Wild violets care – How to grow wild violet plants.* Gardening Know-How. https://www.gardeningknowhow.com/ornamental/bulbs/violet/wild-violets-care.htm#:~:text=While%20they%20tolerate%20many%20soil

Tips to harvest sunflowers. (2020, May 2). Grainvest Group. https://grainvest.co.za/2020/05/02/tips-to-harvest-sunflowers/#:~:text=Harvest%20sunflowers%20when%20their%20petals

Tirrell, R. (1974, June 2). Herb, of folklore. *The New York Times*, 149. https://www.nytimes.com/1974/06/02/archives/herb-of-folklore.html#:~:text=COMFREY%2C%20a%20magical%20herb%20of

Tutorial: How to make a traditional tincture with roots. (2013, October 31). Minnesota Herbalist. https://minnesotaherbalist.com/2013/10/31/tutorial-how-to-make-a-traditional-tincture-with-roots/

Tyler, S. (2019, May 27). *Western larch - New uses for an ancient medicine.* Botanical Medicine. https://www.botanicalmedicine.org/western-larch-new-uses-for-ancient-medicine/

Types of pine trees with identification guide, chart and pictures. (n.d.). Leafy Place. Retrieved March 2, 2022, from https://leafyplace.com/types-of-pine-trees-identification-and-pictures/

Undlin, S. (2020, July 27). *Top mugwort uses (and where to find it).* Plantsnap. https://www.plantsnap.com/blog/top-mugwort-uses-and-where-to-find-it/#:~:text=Identifying%20Mugwort%20Out%20In%20The%20World&text=Its%20dark%20green%20leaves%20are

Valerian. (2015, May 12). Kaiser Permanente. https://wa.kaiserpermanente.org/kbase/topic.jhtml?docId=hn-2179004

Valerian: Valeriana officinalis. (n.d.-a). Edible Wild Food. Retrieved March 4, 2022, from https://www.ediblewildfood.com/valerian.aspx#:~:text=Valerian%20is%20a%20perennial%20plant

Valerian: Valeriana officinalis. (n.d.-b). Edible Wild Food. Retrieved March 4, 2022, from https://www.ediblewildfood.com/valerian.aspx

Verma, R., Gangrade, T., Ghulaxe, C., & Punasiya, R. (2014). *Rubus fruticosus (blackberry) use as an herbal medicine.* Pharmacognosy Reviews, 8(16), 101. https://doi.org/10.4103/0973-7847.134239

Viburnum opulus. (n.d.). Gardeners World. Retrieved February 15, 2022, from https://www.gardenersworld.com/plants/viburnum-opulus/

Viburnum opulus: Cramp bark. (n.d.). Gaia Herbs. Retrieved February 15, 2022, from https://www.gaiaherbs.com/blogs/herbs/cramp-bark

Ware, M. (2019, November 1). *What to know about cranberries.* Medical News Today. https://www.medicalnewstoday.com/articles/269142#:~:text=Many%20people%20consider%20cranberries%20to

Westover, J. (n.d.). *How to identify a purple coneflower when not in bloom.* SF Gate. Retrieved February 22, 2022, from https://homeguides.sfgate.com/identify-purple-coneflower-not-bloom-62343.html

What happens in your body if you drink larch tea. (2021, February 21). Newsy Today. https://newsy-today.com/what-happens-in-your-body-if-you-drink-larch-tea/

What is awl shaped? (n.d.). Movie Cultists. Retrieved February 13, 2022, from https://moviecultists.com/what-is-awl-shaped

White oak. (n.d.). Natural Resource Stewardship. Retrieved March 2, 2022, from https://naturalresources.extension.iastate.edu/forestry/iowa_trees/trees/white_oak.html#:~:text=
White%20oak%20leaves%20are%20simple

Wild foraging: How to identify, harvest, store and use horsetail. (2016, May 2). The Daring Gourmet. https://www.daringgourmet.com/wild-foraging-how-to-identify-harvest-store-and-use-horsetail/#:~:text=What%20-
does%20horsetail%20look%20like,the%20nodes%E2%80%9D%20(Wikipedia).

Wild garlic or ramsons – A foraging guide to its food, medicine and other uses. (n.d.). Eat Weeds. Retrieved February 23, 2022, from https://www.eatweeds.co.uk/wild-garlic-allium-ursinum

Wild horse tail. (n.d.). Recipes from the Wild. Retrieved February 28, 2022, from https://recipesfromthewild.wordpress.com/wild-horse-tail/#:~:text=Creating%20a%20Tincture%3A&text=With-
in%20one%20to%20two%20hours

Wild sarsaparilla (aralia nudicaulis). (n.d.). MPG North. Retrieved March 4, 2022, from https://mpgnorth.com/field-guide/araliaceae/wild-sarsaparilla

Wild violet vinegar infusion. (n.d.). Grow a Good Life. Retrieved March 4, 2022, from https://growagoodlife.com/wild-violet-vinegar/

Wildflowers of the Adirondacks: Wild sarsaparilla (aralia nudicaulis). (n.d.). Wild Adirondacks. Retrieved March 4, 2022, from https://wildadirondacks.org/adirondack-wildflowers-wild-sarsaparilla-aralia-nudicaulis.html#:~:text=Identification%20of%20Wild%20Sarsaparilla

Wildflowers of the Adirondacks: Wintergreen (gaultheria procumbens). (n.d.). Wild Adirondacks. Retrieved March 4, 2022, from https://wildadirondacks.org/adirondack-wildflowers-wintergreen-gaultheria-procumbens.html#:~:text=Identification%20of%20Wintergreen

Wildlife friendly landscapes. (n.d.). NC State Extension. Retrieved February 24, 2022, from https://wildlifefriendlylandscapes.ces.ncsu.edu/#:~:text=Identification%3A%20Japanese%20Honeysuckle%20is%20an

Wilen, C A, et al. *Pesticides: Safe and Effective Use in the Home and Landscape.* University of California Agriculture and Natural Resources, 2019, ipm.u-canr.edu/PMG/PESTNOTES/pn74126.html. Accessed 14 Mar. 2022.

Willis, E. (n.d.). *Herb article - Catnip.* Rebecca's Herbal Apothecary. Retrieved March 2, 2022, from https://www.rebeccasherbs.com/pages/herb-article-br-catnip#:~:text=Preparations%20%26%20Applications&text=Catnip%20can%20be%20prepared%20as

Willow bark. (2021, June 11). Rx List. https://www.rxlist.com/willow_bark/supplements.htm

Willow tree mythology and folklore. (n.d.). Trees for Life. Retrieved March 4, 2022, from https://treesforlife.org.uk/into-the-forest/trees-plants-animals/trees/willow/willow-mythology-and-folklore/

Windling, T. (2016, May 12). *The folklore of nettles.* Myth & Moor. https://www.terriwindling.com/blog/2016/05/from-the-archives-picking-nettles.html

Wintergreen. (2021, June 11). Rx List. https://www.rxlist.com/wintergreen/supplements.htm

Wong, C. (2021a, January 9). *What you need to know about chickweed.* Very Well Health. https://www.verywellhealth.com/chickweed-what-should-i-know-about-it-89437

Wong, C. (2021b, March 3). *What is cramp bark (viburnum)?* Very Well Health. https://www.verywellhealth.com/the-benefits-of-viburnum-cramp-bark-88657

Wong, C. (2021c, November 9). *What is red clover?* Very Well Health. https://www.verywellhealth.com/the-benefits-of-red-clover-89577#:~:text=In%20herbal%20medicine%2C%20red%20clover

Wresting burdock seed and its medicinal uses. (2017, December 1). Northeast School of Botanical Medicine. https://7song.com/wresting-burdock-seed-and-its-medicinal-uses/

Yarrow. (2021, June 11). Rx List. https://www.rxlist.com/yarrow/supplements.htm

IMAGES

Chai, S. (2021). *Anonymous person with bag of plastic bottles* [Online image]. In Pexels. https://www.pexels.com/photo/anonymous-person-with-bag-of-plastic-bottles-7262933/

Cottonbro. (2020). *Leaves and twigs on a canvas bag* [Online image]. In Pexels. https://www.pexels.com/photo/leaves-and-twigs-on-a-canvas-bag-6033829/

Fotios, L. (2018). *Person digging on soil using garden shovel* [Online image]. In

Pexels. https://www.pexels.com/photo/person-digging-on-soil-using-garden-shovel-1301856/

Hamra, J. (2018). *Person Holding Round Framed Mirror Near Tree at Daytime* [Online image]. In Pexels. https://www.pexels.com/photo/person-holding-round-framed-mirror-near-tree-at-daytime-979927/

Hatchett, M. (2019). *Folding knife* [Online image]. In Pexels. https://www.pexels.com/photo/folding-knife-2599276/

Jameson, L. (2021). *Man in blue t-shirt and brown shorts sitting on ground With green plants* [Online image]. In Pexels. https://www.pexels.com/photo/man-in-blue-t-shirt-and-brown-shorts-sitting-on-ground-with-green-plants-9324354/

Jess Bailey Designs. (2017). *Close-up photography scissors* [Online image]. In Pexels. https://www.pexels.com/photo/close-up-photography-scissors-755991/

Kool, A. (2015). *El Capitan on a sunny afternoon* [Online image]. In Unsplash. https://unsplash.com/photos/ndN00KmbJ1c?modal=
%7B%22tag%22%3A%22Login%22%2C%22value%22%3A%7B%22-
tag%22%3A%22Like%22%2C
%22value%22%3A%7B%22photoId%22%3A
%22ndN00KmbJ1c%22%2C%22userId%22%3A
%22AjxdOkQCJr8%22%7D%7D%7D

Lewis, J. (2020). *Yellow leaf in close up photography* [Online image]. In Pexels. https://www.pexels.com/photo/yellow-leaf-in-close-up-photography-5497873/

Mazumder, A. (2018). *Person holding a green plant* [Online image]. In Pexels. https://www.pexels.com/photo/person-holding-a-green-plant-1072824/

Monstera. (2021). *Wooden brushes prepared for washing and cleaning* [Online image]. In Pexels. https://www.pexels.com/photo/wooden-brushes-prepared-for-washing-and-cleaning-6621326/

Patel, S. (2020). *Green leaves on tree branch* [Online image]. In Pexels. https://www.pexels.com/photo/green-leaves-on-tree-branch-4400283/

Pixabay. (2012). *Green Leaf Plant* [Online image]. In Pexels. https://www.pexels.com/photo/green-leaf-plant-86397/

Pixabay. (2016). *Green tree plant leaves* [Online image]. In Pexels. https://www.pexels.com/photo/green-tree-plant-leaves-40896/

Rodnikova, M. (2021). *Pruning shears and a flower wreath* [Online image]. In Pexels. https://www.pexels.com/photo/pruning-shears-and-a-flower-wreath-9797607/

Schwartz, K. (2021). *Green leaves of growing fern* [Online image]. In Pexels. https://www.pexels.com/photo/green-leaves-of-growing-fern-8117889/

Sunsetoned. (2020). *Crop woman reading book on grass* [Online image]. In Pexels. https://www.pexels.com/photo/crop-woman-reading-book-on-grass-5981113/

Tis, A. (2021). *Yellow maple leaf on yellow background* [Online image]. In Pexels. https://www.pexels.com/photo/yellow-maple-leaf-on-yellow-back-ground-9563330/

Tran, V. (2019). *Photo of jar near cinnamon sticks* [Online image]. In Pexels. https://www.pexels.com/photo/photo-of-jar-near-cinnamon-sticks-3273989/

Webb, S. (2018). *Green leafed indoor plant* [Online image]. In Pexels. https://www.pexels.com/photo/green-leafed-indoor-plant-1048035/

www.ingramcontent.com/pod-product-compliance
Lightning Source LLC
Chambersburg PA
CBHW022112210326
41597CB00047B/212